Management of Historic Centres

Conservation of the European Built Heritage Series

Series editor: Robert Pickard

A series of books examining a wide range of issues in conservation of the built heritage, published in association with the Council of Europe

1 Policy and Law in Heritage Conservation
Edited by Robert Pickard

2 Management of Historic Centres
Edited by Robert Pickard

Forthcoming:

3 Financing the Preservation of the Architectural Heritage
Edited by Robert Pickard

Management of Historic Centres

Edited by
Robert Pickard

London and New York

First published 2001
by Spon Press
11 New Fetter Lane, London EC4P 4EE

Simultaneously published in the USA and Canada
by Spon Press
29 West 35th Street, New York, NY 10001

Spon Press is an imprint of the Taylor & Francis Group

Typeset in Times New Roman by
Exe Valley Dataset Ltd, Exeter, Devon
Printed and bound in Great Britain by
Bell & Bain Ltd, Glasgow

British Library Cataloguing in Publication Data
A catalogue record for this book is available from the British Library

Library of Congress Cataloging in Publication Data
Management of historic centres / edited by Robert Pickard.
p. cm – (Conservation of the European built heritage series)
Includes bibliographical references and index.
1. Urban policy – Europe. 2. Cities and towns – Europe.
3. Historic sites – Europe. 4. World Heritage areas – Europe.
5. Cultural property – Protection – Europe. I. Pickard, Robert. II. Series.

HT169.E85 M35 2000
307.76'094—dc21 00–044058

ISBN 0-419-23290-7

Contents

Contributors

Brigitte Beernaert trained as an architectural historian in 1974 and subsequently was awarded a Degree in Communication Sciences from the University of Ghent, Belgium, in 1975. Since 1977 she has been a member of staff at the Department of Historical Monuments and Urban Renewal of the City of Bruges. In the period 1977–88 she was closely involved in the realization of Section Plans. Since 1988 she has been responsible for the organization of the Architectural Heritage Days in Bruges. She has also been involved in research work on the history of monuments and historic houses for restoration purposes and has been the author of several publications on the history of architecture and restoration in Bruges.

Silvia Brüggemann read chemistry in Münster/Westf (1980–83) and history and history of art in Bochum (1984–92). In 1989 she was Magistra Artium of the College of Ruhr–University Bochum and in 1989/90 practised as an architect. In 1992/3 she worked for the Foundation of Weimar Classic. Since then she has carried out private research on building history and in 1995 she founded the 'Büro für Baugeschichte' (private research office on building history) (co-owned with Christoph Schwarzkopf since 1999). She has lectured on building history at Fachhochschule Erfurt between 1994 and 1997 and is the author of various articles on conservation issues.

Bruno Coussy is an architect and town planner, and is Director of the PONAT Stratégies Urbaines Agency in Rochefort, France. As an advisor in town planning for various local authorities, he has carried out extensive research with the *Direction de l'Aménagement et de l'Urbanisme* (Town and Country Planning Authority) in the field of urban forms and sustainable development in towns, on the subject of urban diagnostics, on the relation between the climate and planning and on regulatory tools for protecting the built heritage. He is also an expert consultant to the Council of Europe's Technical Co-operation and Consultancy Programme for the integrated conservation of the cultural heritage.

Nathaniel Cutajar studied history and archaeology at the University of Malta. He subsequently continued his studies at the University of York (UK), obtaining an MA in Archaeological Heritage Management in 1994. Since 1989 he has worked in various capacities with the Museums Department (Malta) and in 1996 he was appointed to the post of Curator responsible for the Heritage Information Management Unit. His particular area of interest includes the study of the medieval archaeological heritage of the Central Mediterranean region.

Juris Dambis graduated from Riga Politechnical Institute, Department of Architecture in 1979 and completed a special course in restoration in 1980. From 1979 to 1989 he was Chief Architect at the Ministry of Culture of the Republic of Latvia. Since 1989 he has been Head of the State Inspection for Heritage Protection. During this period he has developed the concept of cultural heritage preservation and has been responsible for drafting legislation in this field, subsequently approved by the Parliament of Latvia. He is a Deputy of the Council of the Latvian Institute and a member of the Architects Society Board of Latvia. From 1993 he has been actively involved in the work of the Cultural Heritage Committee of the Council of Europe and has chaired this committee since 1999. Other areas of work have included participation in international conferences, seminars, workshops in Europe and the implementation of several international projects regarding heritage protection in Latvia.

Werner Desimpelaere is qualified in architecture and town planning and is the Senior Partner and President of the Groep Planning – a multidisciplinary professional partnership based in Bruges and Brussels, Belgium, specializing in the fields of regional and town planning, traffic planning and architecture. He has lectured and participated in many symposiums on town renewal, rehabilitation and the integration of contemporary architecture in historic cities in several European and South American countries. He has acted as an expert consultant to the Cultural Heritage Department of the Council of Europe since 1991 and to UNESCO since 1996.

Miloš Drdácký was awarded a Diploma Engineering Degree from the Czech University of Technology in Prague in 1968, and a scientific degree, CSc (corresponding to PhD), from the Czechoslovak Academy of Sciences in 1978. He is the Director of the Institute of Theoretical and Applied Mechanics of the Academy of Sciences where he has been employed as

Chief Research Fellow since 1968. In 1993 he founded an Associated Research Centre for Historic Structures and Sites and served as its Head until 1998. During the same period he was in charge of the Town Architect Office in the World Heritage City of Telč (Moravia) and co-ordinated a programme in association with the Council of Europe's Technical Co-operation and Consultancy Programme in the field of cultural heritage to develop the 'Telč XXI Centre'. He has been an external lecturer at the Czech Technical University in Prague (1975–1980), the University of Industrial Arts in Prague (1984–1990), and the Technical University in Liberec – Faculty of Architecture and Technical University in Brno – since 1998. He is a member of IABSE, IASS, ISTL and several Czech professional associations, and member of a number of editorial boards. His major research activity in recent years has focused on the performance of historic structures and environments, their degradation and safeguarding, on the documentation of cultural heritage and the management of historic cities. He has authored more than 160 publications and edited various proceedings and books in the above fields.

Xerardo Estévez was born in Santiago de Compostela. He trained as an architect at the School of Barcelona, Spain. He was Mayor of Santiago de Compostela between 1983 and 1986 and from 1987 to 1998. His experience as an architect in the government of a historic city allowed him the opportunity to explore, by means of political involvement, the profession's potentiality for actively participating in urban development.

Giorgio Gianighian is a graduate in architecture from the Istituto Universitario di Architettura di Venezia, Italy (IUAV) (1970). Since 1973 he has held various teaching and research positions in conservation and restoration at the IUAV and has been Visiting Professor at various universities in Canada, Japan, Israel and the United Kingdom. He has lectured and participated in many international conferences on conservation and urban restoration. In 1997 he became a Fellow of the Japan Society for the Promotion of Science in Japan. In 1998 he was responsible for the PVS-IUAV working unit for the survey of a historic area in Mostar (Bosnia-Herzegovina), via the Aga Khan Trust for Culture and for the World Monuments Fund. He has an international publication record and runs a studio/practice in Venice.

Kakha Khimshiashvili trained as a restoration architect at the Department of Restoration, Architecture Faculty, Tbilisi State Acadamy of Fine Arts, Georgia, between 1978 and 1984, and subsequently undertook postgraduate doctoral research on the History and Theory of Architecture from 1985 to 1988. Between 1994 and 1995 he studied at the University of York (UK) and was awarded an MA in Conservation Studies. He is a member of the ICOMOS Georgia National Committee. He has held various academic and technical positions in the field of architectural restoration including for the Academy of Science of the Georgian Republic, the Tbilisi State Academy of Fine Arts, the Main Board for Monuments Protection, and is currently Technical Co-ordinator of the Fund for the Preservation of the Cultural Heritage of Georgia.

David Lovie has a Degree in Land Use Studies, and a Masters Degree in Urban Planning Design – spending 5 years as a research associate in Urban Planning Design at the University of Newcastle upon Tyne. In 1972 he started a career as a Conservation Officer in local government during which time he also created and chaired an independent urban studies organization; chaired the Royal Town Planning Institute National Environment Education Panel for 2 years; and initiated and supported a series of twenty-five regional BBC television programmes on townscape. In 1986 he became Newcastle's City Conservation Officer and was subsequently seconded to the Grainger Town Partnership as Heritage Officer in 1996. In 1997 he chaired the Institute Steering Committee of the Association of Conservation Officers and was responsible for over-seeing the establishment of the Institute of Historic Building Conserv-ation in 1998 as a professional body in the United Kingdom. He has also been actively involved in education concerning the historic built environment for over 25 years as animateur, lecturer and university external examiner.

Elene Negussie was awarded a Bachelor and Masters Degrees of Social Science with a major in Human Geography, from Stockholm University in 1997. The Degrees involved studies at Trinity College, Dublin and York University, Canada. Since 1998 she been involved in a doctoral research programme in the Geography Department, Trinity College, Dublin. The research explores different approaches to urban conser-vation with a qualitative and longitudinal study of temporal and spatial change in the cities of Dublin and Stockholm, addressing how social factors and trends determine what is worth conserving in the urban built environment.

Anthony Pace read history and archaeology at the Universities of Malta and Cambridge. He is currently Director of the Museums Department which is responsible for national museums, the management of Malta's World Heritage and archaeological sites as well as providing super-intendence services in heritage management. His research interests lie mainly in Mediterranean prehistory, museology and policy development in the heritage sector.

Robert Pickard completed his first Degree in surveying in 1979 and became a Professional Associate of the Royal Institution of Chartered Surveyors (RICS) in the United Kingdom in 1983. He was awarded a PhD in planning law and practice in 1990 and a postgraduate qualification in building conservation in 1992. In 1993 and 1998 he received research awards from the RICS Education Trust to investigate European Policy and Law in the field of conservation and to examine Methods of Funding the Preservation of the Architectural Heritage in Europe and North America respectively. Since 1994 he has been an expert consultant to the Council of Europe's Technical Co-operation and Consultancy Programme in the field of cultural heritage and in 1997 became a member of its Legislative Support Task Force (which he co-ordinated on secondment to the Council of Europe between 1998 and 2000). He was a member of the Education Committee of the Association of Conservation Officers between 1995 and 1998 (subsequently the Institute of Historic Building Conservation – Member since 1998). He has lectured on property investment, town and country planning and conservation issues; edited and authored books and articles and participated in several international conferences and events on conservation issues; and is external examiner to the MA in Historic Conservation at Oxford Brookes University/University of Oxford.

Christoph Schwarzkopf was awarded a Diploma of the College of Architecture and Building in Weimar in 1989 and subsequently practised as an architect. From 1990 to 1994 he was Conservation Officer at the Thüringian Land Office for the Preservation of Historical Monuments. In 1994/95 he worked for the Foundation for Thüringian Palaces and Gardens. From 1995 to 1999 he was employed in the Büro für Baugeschichte (private research office on building history). Since 1999 he has been a self-employed architect and co-owner of the Büro für Baugeschichte, together with Silvia Brüggemann. He is a member of ICOMOS and since 1996 he has lectured on the Preservation of Monuments at the Bauhaus University in Weimar. He has written a number of articles on conservations issues.

Erling Sonne qualified as a landscape architect from the School of Architecture at Aarhus, Denmark, in 1979. In 1979, he was employed as a planning architect with the local authority of Hjørring, where, in addition to work on city and landscape planning, he also acquired insights into urban design and preservation issues. In 1987 he became a senior architect and buildings inspector with the local authority of Ribe, where his responsibilities include for planning, urban development, preservation, and building works in Ribe. During recent years, he has devoted a great deal of attention to holistic tasks with a special focus on the interplay between preservation and development in the city centre of Ribe. In addition to this, he has given a substantial number of lectures and has written several articles on this topic.

Mikhäel de Thyse trained in the history of architecture, urban planning and conservation methods at the R. Lemaire Centre for Conservation, Catholic University of Leuven/Louvain, Belgium. Following this, he set up a private consultancy office specializing in built heritage issues in Belgium. He joined the Cultural Heritage Department of the Council of Europe in 1992. He is currently the programme administrator for the Technical Co-operation and Consultancy Programme for the integrated conservation of the cultural heritage, responsible for the implementation of international projects mainly in countries within central and eastern Europe. This includes the organization of technical workshops and expert assistance for the development of pilot projects, and the co-ordination of studies by *ad hoc* groups of experts on inventory methods, rehabilitation and the interpretation of the heritage.

Foreword

The first book in this series on the Conservation of the European Built Heritage examined issues of 'Policy and Law in Heritage Conservation' through consideration of themes drawn from the guiding principles established by the Granada and Malta Conventions – the Council of Europe's conventions for the protection of the architectural and archaeological heritage respectively.

This second book scrutinizes the strategies that have been adopted for integrated action with respect to the 'Management of Historic Centres' through examination of management policies and associated planning mechanisms, regeneration action and inter-related issues concerned with the environment, tourism and heritage protection in twelve historic centres in Europe.

Effective action is needed to promote the rehabilitation and management of historic centres in Europe and, in this context, to rationalize the past with the present and future needs of communities. The approach to be adopted must ensure that conservation of the built heritage, regardless of its monumental value, includes a framework for urban projects and a system of urban management that is sustainable by assisting social cohesion and economic vitality and improving the quality of life for inhabitants. This study highlights the current situation and progress made in a diverse range of small and large cities, and centres of larger cities, spread throughout Europe.

This examination of the approaches used to tackle the complex issues of preserving, managing and enhancing historic centres makes an important contribution to the Council of Europe's 1999–2000 campaign: 'Europe, A Common Heritage'.

José María Ballester
Head of the Cultural Heritage Department
Directorate of Culture and Cultural Heritage
Council of Europe

Robert Pickard

Introduction

Context of the study

At the Second Summit of the Council of Europe held in Strasbourg in October 1997, the Heads of State and Governments of the member states of the Council of Europe reaffirmed the importance attached to 'the protection of our European cultural and natural heritage and to the promotion of awareness of this heritage'. An Action Plan adopted by the summit decided on the need to launch a campaign on the theme *Europe, A Common Heritage* to be held 'respecting cultural diversity, based on existing or prospective partnerships between government, education and cultural institutions, and industry' (Fig. 1.1). This campaign, which was launched in September 1999 to last for one year, takes place 25 years after the European Architectural Heritage Year (1975).

The 1975 campaign marked the start of the Council of Europe's activities that gave rise to the *European Charter of the Architectural Heritage*. The charter identified the fact that the structure of historic centres is conducive to a harmonious social balance and that by offering the right conditions for the development of a range of activities our old towns can be helped to become a favourable environment for social integration. The subsequent *Amsterdam Declaration* of the Congress on the European Architectural Heritage (Council of Europe, 1975) which introduced the concept of 'integrated heritage conservation' emphasized that all areas of towns form part of the architectural heritage and that there is a responsibility to protect them against the threats posed by neglect, deliberate demolition, incongrous new construction and excessive traffic. It further stated that the rehabilitation of old areas should be carried out in a way which ensures that there is no need for a major change in the social composition of the residents. Integrated conservation involves an acknowledgement of their responsibility by local authorities, the participation of citizens, legislative reform to ensure

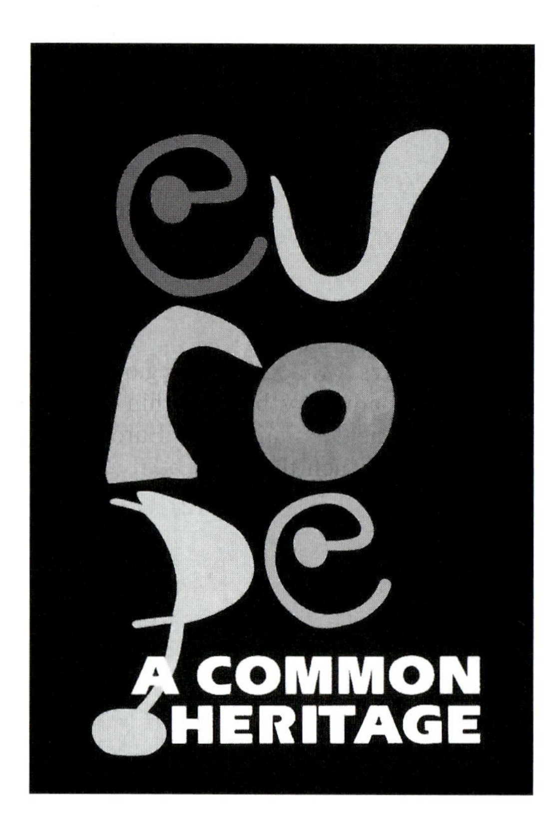

Fig. 1.1 Europe – A Common Heritage: a Council of Europe Campaign 1999–2000 (campaign logo).

action to safeguard both individual buildings and architectural complexes in the urban context, administrative measures to plan and co-ordinate actions, and appropriate financial mechanisms to support such action.

The concept of integration is now enshrined in two founding texts. The contracting parties to the Convention for the Protection of the Architectural Heritage of 1985 (the Granada Convention) (Council of Europe, 1985) have undertaken to take statutory measures to protect the architectural heritage, satisfying certain minimum conditions laid down in the Convention. These include the maintenance of inventories, the adoption of integrated conservation policies and the establishment of the machinery required for consultation and co-operation in the various stages of the decision-making process (including both cultural associations and the public), the provision of financial support and the fostering of sponsorship and non-profit making associations (the latter has been further explored in Council of Europe, 1991). The European Convention on the Protection of the Archaeological Heritage (revised) (1992) (the *Malta Convention*) (Council of Europe, 1992) took note of the fact that the growth of major urban development projects made it necessary to find ways of protecting this heritage through integrated conservation methods.

Since 1975 the Council of Europe has been actively involved in promoting the concept of integrated conservation. More recently the Cultural Heritage Department's *Technical Cooperation and Consultancy Programme for the integrated conservation of the cultural heritage* has provided technical assistance for urban pilot projects in several countries, most recently to help the new member states of central and eastern Europe. Since 1998 it has run a series of workshops on 'rehabilitation in old centres as a factor of social cohesion and economic development' and it is planned to hold a conference on this theme in Lisbon, Portugal, in May 2000.

The Congress of Local and Regional Authorities of Europe (CLRAE), which is an advisory body within the institutions of the Council of Europe, has held a number of European Symposia on Historic Towns since 1971 in which there has been an exchange of experiences between the authorities of the historic towns of Europe. Moreover, in the context of the campaign *Europe, A Common Heritage* the CLRAE is responsible for the organization of a transnational project to establish a European Association of Historic Towns. The aims of this project include: to assist in the establishment of national associations of historic towns, to share experiences of good practice on urban conservation and management, to promote viability and sustainability, to encourage partnership and cooperation between historic towns and collaboration with institutions, to raise public awareness and uphold the principles of democratic participation in the management of historic cities, and to promote the development of appropriate national legislation which protects and enhances the historic heritage.

Other international bodies have also emphasized the importance of the conservation in historic centres. In 1976 UNESCO recommended that the necessary steps should be taken to ensure the protection and restoration of historic towns and areas and to ensure their development and harmonious adaptation to contemporary life (UNESCO, 1976). Furthermore, the ICOMOS *Washington Charter* states that the conservation of historic towns and urban areas 'should be an integral part of coherent policies of economic and social development and of urban and regional policy at every level' (ICOMOS, 1987).

Themes of the study

Within this context the aim of this study is to examine the following key themes for the management of historic urban centres within a representative sample of centres in different European countries. The twelve historic centres that have been chosen are spread throughout Europe, each one from a separate country. They are diverse in character and the range includes small towns, cities and urban centres within cities. Some of the centres have been designated by UNESCO as World Heritage Sites/Cities, or have applied for such designation, whilst others are recognized as European Cities of Culture or are seeking to be chosen as such. The centres have all faced different problems and a variety of management approaches have been (or are being) utilized and will be the subject of examination. For some centres the process of developing a management strategy is at an early stage, in others it is well advanced.

The campaign *Europe, A Common Heritage* provides the appropriate vehicle and timing for this study.

The policy and planning framework

The recording of the condition of the immovable heritage within historic centres, areas and quarters can be a useful starting point for co-ordinating action to revitalize the heritage and to recognize its potential not only in terms of cultural value but also in terms of functional, economic or other values. Some countries have established specific programmes to identify the relative condition and the occupancy and vacancy position of architectural monuments as a starting point for more concerted programmes of action for the conservation, repair, restoration, regeneration and management of historic environments. For those centres initiating such strategies this will be an important part in the process.

The *Amsterdam Declaration* identified that 'integrated conservation necessitates the adoption of legislative and administrative measures' and the need for co-ordination at different levels. The ICOMOS *Washington Charter* further identified the need for 'multidisciplinary studies' on historic centres. It stated that 'resultant' conservation plans should address all relevant factors including archaeology, techniques, sociology and economics. The extent to which this has been achieved will provide an insight into the progress made in individual centres. The idea of introducing conservation management plans may be in the early stage of development but there should be some form of administrative co-ordination for effective historic centre management.

As a minimum there is likely to be some method to define an area and to designate monuments, groups of buildings, sites and areas within a centre, including urban complexes or even separate historic districts. Consideration of policies to define the centre may be identified as well as the system for monitoring demolition, large-scale alterations and new construction work. Moreover, the idea of managing 'change' within an area will be an important step in the process of keeping an area alive.

It is likely that management schemes will be integrated within a wider framework of urban planning policies to ensure the protection of the heritage (as recommended by both the Granada and Malta Conventions). In this context the following issues may also be relevant.

- Policies on applications for new development including the circumstances in which it may not be permitted; the compatibility of new functions within the area; respect for the historic context in terms of volume, scale, form, materials and quality of design.
- Specific safeguards to protect views, vistas and settings, and historic street layouts.

- Control procedures and sanctions and coercive measures to safeguard the built fabric and other areas of recognized importance (including sites of archaeological importance, open spaces and historic parks and gardens).
- Strategies to enhance the appearance and character of the area through the use of design guides and development briefs for sites that have may been regarded as having a negative or neutral impact within a centre.

Management and regeneration action

The management of a historic centre will require the formulation and implementation of some form of plan mechanism and may include a specific 'conservation plan' or 'action plan' and the use of an economic development and regeneration strategy to encourage the maintenance and re-use of historic buildings and environmental improvements. The improvement or rehabilitation of buildings for housing, and the encouragement of compatible businesses, are likely to be two basic objectives of conservation as they may help to ensure that an area is kept alive. In this respect, action taken to preserve single monuments as well as groups and sites will be equally important. This will require some consideration of the levels of vacancy within heritage assets, and an assessment of the whether historic buildings are sufficiently flexible to accommodate new uses without damaging their essential character.

National and local authorities and other agencies may need to take part in the management process and this may involve the establishment of specific management agencies (public, private and joint venture) for the co-ordination of policies to revitalize a historic centre. Housing rehabilitation organizations, commercial organizations and voluntary bodies may also have a role to play in this context.

Specific financial arrangements will be needed to encourage preservation action, environmental improvements and regeneration activity, including funding partnerships or joint venture arrangements between the public authorities and private organizations perhaps their involvement of foundations. Such arrangements maybe more directly linked to urban renewal and regeneration organizations and mechanisms.

Environmental management

Comprehensive environmental erosion caused by heavy traffic and inappropriate traffic management measures, and other pollutants, can have a significant impact, not just on individual buildings, but also on the whole environment of a historic centre. It will be important to consider opportunities for reducing pollution and vibration, to remove traffic and reduce congestion, and to improve the physical fabric of the environment including road surfaces, street furniture and other features such as landscaping and open spaces, and by the use of pedestrian priority schemes and the planning of parking areas.

Thus, it can be seen that the co-ordination of the authorities responsible for the construction of new roads, other transport facilities and a range of environmental protection issues will play an important role in safeguarding a historic centre.

Tourism and heritage management
Tourism can prove to be economically beneficial to historic centres. It can create positive gains for hotels, shops and other businesses and it can positively support conservation action. It also needs to be balanced against the needs and wishes of the local population. An over-emphasis on supporting tourist activity may damage the balance of existing communities and damage cultural assets themselves. Tourism can create pressure for new services and associated development. The extent to which tourist activity can be supported needs to be carefully planned and controlled. Heritage management is also important for maintaining the cultural identity of historic centres.

Sustainability
Taking into account the foregoing themes, the question of sustainability arises. The aim today should be to ensure that the management of historic centres is sustainable in terms of utilizing and safeguarding heritage assets for future generations through rehabilitation, maintaining a social balance and employment opportunities, environmental considerations, the management of tourist activity and participation by he community. Integrated conservation approaches encourage this process. Moreover, in some countries policies for 'sustainable development' have now been developed in relation to land-use planning and heritage management.

The extent to which this approach is now filtering through to particular management strategies in historic centres is uncertain. In some cases, specific policies will have been implemented, in others sustainable goals may be implicit or at a preliminary stage. One aspect of this is the realization that not only is it important to preserve cultural assets *per se*, but also whether the capacity to allow change within the historic centre can be preserved. Change within a historic environment may often be the key to long-term preservation and sustainability.

Summary
Through these themes the present study seeks to provide evidence of different approaches to the management of historic centres within Europe. The analyses of a number of individual centres and a review of the key principles for historic centre conservation and rehabilitation will provide an opportunity to reflect on and assess the types of management tools and approaches that are required in order to sustain the quality of historic centres as assets of the 'common heritage'.

References

Council of Europe (1975): *Amsterdam Declaration*: Congress on the European Architectural heritage, 21–25 October 1975.

Council of Europe (1985): *Convention for the Protection of the Architectural Heritage of Europe (3 October 1985)* (ETS, No. 121) (Granada Convention).

Council of Europe (1991): *Funding the Architectural Heritage.* (Published with the collaboration of the Association of Italian Savings Banks and the National Centre of French Savings Banks).

Council of Europe (1992): *European Convention on the Protection of the Archaeological Heritage (revised)* (16 January 1992) (ETS, No. 143) (Malta Convention).

ICOMOS (1987): *Charter for the Conservation of Historic Towns and Urban Areas* (the *Washington Charter*).

UNESCO (1976): *Recommendation Concerning the Safeguarding and Contemporary Role of Historic Areas* (Warsaw and Nairobi, 1976).

Brigitte Beernaert and Werner Desimpelaere

Bruges, Belgium

Introduction

Bruges lies 13 km from the North Sea in the west of Belgium and is the provincial capital of West Flanders. Greater-Bruges was formed by the fusion of the old city with seven surrounding districts and comprises 13,100 hectares, whilst the medieval city is just 410 hectares. At the time of writing Bruges has 115,575 inhabitants, of which 20,275 reside in the old city.

Bruges has retained its medieval streetplan and this, coupled with the small-scale city and its canals, constitute her most important characteristics. The city's architectural heritage ranges over diverse periods with a surprising number of authentic medieval buildings and fabric preserved. Constructions from later periods nearly always adapted to the medieval context in their scale, rhythm and proportions, and old elements were frequently re-used or incorporated.

It is worth noting that the nineteenth century had a great impact on the medieval aspect of Bruges, resulting in its present-day fame, it is often the reason why critics dismiss the city as a fake or pastiche today.

Between 1200 and 1500, Bruges was one of the economic capitals of north-west Europe and as a result was an extremely thriving place due to the commerce of its harbour and proximity to the sea. The old city's boundary is still formed today by the large embankment of 1297. This embankment was fortified in the fourteenth and fifteenth centuries by the addition of nine city gates, four of which survive. Roads lead from the gates to the centre of the city and remain the most important commercial streets.

The *Dijver* (etymologically from the Celtic, meaning *holy water*) is now one of the most important tourist thoroughfares. This part of the town was certainly inhabited by the beginning of the Christian Calendar; and in the north of the city a Gallo-Roman settlement was present until circa 270. Archeological excavations suggest that several small settlements were placed along the old Roman road between Oudenburg and Aardenburg, some of which became villages or towns, Bruges being one of them.

In the ninth century the town *Bryggja* developed around the Burg, where the Count of Flanders had his residence, and the Oude Burg, where the earliest rich cloth merchants were established. The Market, in immediate proximity to the Burg, was the place where municipal autonomy was displayed, the burghers meeting here on important occasions. The imposing thirteenth-century belfry with its halls was a symbol of this autonomy and economic wealth.

Material wealth led to an upsurge in building: churches, monasteries, a beguinage, and hospitals rose skywards as witnesses to the city's religious and social convictions. Mansions and small houses were built side by side and during the height of the building period – i.e. the fourteenth century – 45,000 inhabitants resided inside the city. However, a slow disintegration of this medieval metropolis began towards the end of the fifteenth and beginning of the sixteenth century. Political problems, the restrictive commercial system and the silting of the channel to the sea, little by little began to take their toll. Undoubtedly though, the sixteenth century was a notable time for Bruges, a period of intense cultural and artistic growth (Fig. 2.1).

During the seventeenth and eighteenth centuries the city continued its efforts to solve the silting problem in the hope of recovering its past economic position, but without success. Architecturally speaking, interesting projects were still being realized; many private homes from this period date from this period.

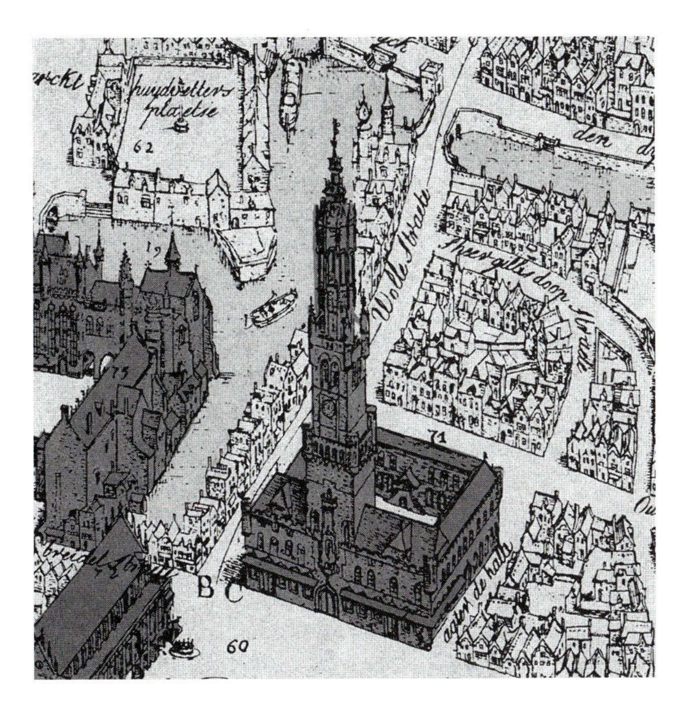

Fig. 2.1 The Belfry was the symbol of municipal autonomy. Town plan of Marcus Gerards, 1562. The Belfry and Cloth Hall dominate the Market Place to this day.

By the second half of the nineteenth century new plans for a harbour connecting Bruges to the sea, were proposed and finally accepted. Work started during the reign of Leopold II (1865–1909) and Bruges' new sea-harbour, Zeebrugge, was inaugurated in 1907. Important to this same period was the architectural revival movement. Gothic architecture was the inspiration for both restoration and new projects. The city itself subsidized *artistic restorations* from 1877. There was new hope for ancient Bruges; a hope which in fact never materialized due to the two world wars. Bruges, however, luckily escaped from the general devastation.

The economic situation after the wars was anything but bright. Work opportunities were minimal and living conditions in the old inner city were lamentable, although help was around the corner with the expansion of industry in Western Europe. This was reflected in a relatively large number of factories appearing on the outskirts of Bruges and originally attracted by the harbour facilities of Zeebrugge.

In the meantime, the city's appearance had degenerated and the carefully planned restoration movement was forgotten. Change was called for, and in 1965 this came via a private initiative – the Marcus Gerards Foundation – whereby a group of concerned inhabitants succeeded in making the local government aware of the actual problems and the possible solutions.

The Department of Historical Monuments and Urban Renewal was established in 1971 in order to guide architectural and building projects and alterations. A city alderman was made responsible for this matter for the first time. For the surrounding district a different Urban Department was installed.

The policy and planning framework

For the last 30 years, an enormous amount of work has been done on renovation, restoration and supervision of the historical centre of Bruges. This is constantly being followed up, guided and controlled.

The Structural Master Plan of 1972

The history of Bruges, like that of all old towns, is a mixture of growth and constant change. A plan was necessary that both guaranteed the conservation of the essential character but could also direct the changes necessary to be compatible with a modern lifestyle.

Bruges is a unique mozaic of small buildings around twisting streets and reflective canals, with private homes as its most fundamental element. The city was originally conceived for pedestrians, with small shops and trading houses, and not for large industrial or commercial concerns. The old buildings became increasingly threatened by traffic, further commercial development and increased prosperity. The political situation was unsympathetic in the struggle between conservation and demolition and, until 1970, there was virtually complete deadlock between the financial and political institutions on the one hand, and the urban development projects on the other. The national government even granted subsidies to demolish run-down properties without considering their historical value.

At the end of the 1960s, the situation in Bruges was as follows:

- an obvious struggle between the old town and the surrounding boroughs;
- functional loss of many buildings in the old city;
- few job opportunities and little industry;
- an impoverished innercity population (the new rich built villas in the suburbs);
- a high proportion of empty houses (one in eight was either empty or neglected);
- many slums;
- the absence of a local urban development policy.

After the 1970 amalgamation of the old inner city with the surrounding suburbs, the new city council was very conscious of the new phenomenon of a complicated urban renewal, and was only too willing to support a strong basic philosophy for the urban development problem. A Structural Plan or Master Plan was designed in 1972 as the best means of supervision at local level (City of Brugge, 1976). At the time this plan was the first of its kind in Belgium. This Master Plan was seen as a thinking model, an organic planning system capable of evolving and adapting to the ever-changing conditions caused by town planning and economic situations. The historical towns would again become promulgators of a *collective memory* without curbing modernization. In contrast they would be concerned with the human and cultural aspects necessary to support economical and technical revolutionary changes. A dynamic approach of such proportions had previously been unthinkable under the prevailing view of town planning.

The Bruges Master Plan influenced not only the town planning policy in Bruges, but caused similar institutions to reflect on their situations throughout Flanders and Belgium. It became a model example, copied, and internationally recognized.

Aims of the Structural Master Plan of 1972

The concept was threefold: the first phase was a survey, the second a detailed examination of each and every quarter of the city and the third stage consisted of feedback on the detailed studies to the previously decided structure. Altogether this format would make continuing adaptation and flexibility possible. It was to be overseen and guided by a steering committee.

The most important aims of this Master Plan were:

- The vitality and appeal of the historic city centre must be guaranteed at all costs, and in order to attract more city dwellers, housing and living conditions should be optimally improved (Tanghe *et al.*, 1984).
- Large-scale businesses, generating any form of nuisance or disturbance, must be removed to the town's boundaries or suburbs. The historical inner city should once again become a typical collection of small-scale inner-town functions for which it was designed; ground use and other

activities with a nuisance factor or atypical small-city occupations should be banned.

- The commercial heart of the city must be injected with new dynamism and made attractive so that equal competition exists between them and out-of-town trade, whether this be shops, cafés or restaurants etc.
- It goes without saying that conservation of the city's historical character requires the preservation of its architectural heritage and townscape. Each new building, be it in scale, volume, shape, rhythm or colour, must conform to the strictest architectural criteria. New buildings of a high quality must be encouraged. The validity and continuation of the historical heritage can only be effective if these buildings are themselves responsible for their own economical viability; it is imperative that they be integrated into the city's lifestyle.
- To improve the character of the inner city, it is absolutely necessary for a strong policy to offset the very real risk of dilapidation. Noise and pollution must be eliminated in order to encourage the return of city dwellers.
- A traffic policy must be designed to solve the problem of conflict between pedestrian and car; the car must be made to conform with pedestrian habits. The traffic plan must be capable of evolving, the philosophy being that it is traffic that must adapt and not the town. To this end, the city has been divided into nine parts, regarded as *rooms*; as in a house, these *rooms* are connected with *corridors* or *traffic-loops*, which conduct traffic from one section of town back to the ring road. Through-traffic is thereby discouraged and redirected to the city's extremities. Public transport is encouraged, and car parks constructed.

This revolutionary – for Belgium – Structural Master Plan was completed by a cost analysis and feasibility study. The anticipated period for the realization of the Plan was 20 years.

The second concept: the section plan

The second part of the Structural Master Plan was launched in 1973. This consisted of an inventory of all buildings as a basis for the section plans proposed for the inner city. The historical city was divided into nine districts, they were then subdivided into fifty blocks, each having a corresponding section plan. Each section plan consisted of ten sub-plans comprising inventory, evaluation and projection.

- The inventory grouped the following data: function of the buildings and traffic directions; roof types, open spaces and street materials; property structures and houses occupied by their owners or tenants; building information on houses with eventual inclusion in the architectural inventory of 1967/75, and officially listed buildings.
- The evaluation comprised: condition of the physical building structure varying from good to urgently in need of improvement with notice being taken of whether the building is abandoned or not; town planning

evaluation of each and every building ranging from good to alarming, plus an evaluation of the façades and the quality of the shopfronts; an architectural-historical survey on a scale from remarkable to neutral; and a tree evaluation.
- Finally, the new projection plan comprised all previously noted problems in the inventory and evaluation plans and possible solutions were advanced.

The section plans were unfortunately never permanently updated or revised due to insufficient personnel at the Department of Historical Monuments and Urban Renewal.

The New Structural Plan for Bruges

The general principles of the original Structural Master Plan are still followed to this day, but after 30 years, the problems have somewhat changed; the radical change brought about in the city's dynamism brought the city council to the decision to commission a new Structural Plan functioning as policy and guideline for further city development.

Most of the aims of the Structural Plan of 1972 have been achieved, but several problems confronting the historical city remain, i.e. the difficulty in protecting the function of city dwellings, an increase in tourism, and also – in relation to the service sector – mobility and conservation and heritage protection.

The options for the historical city centre in this New Structural Plan are as follows:

- Bruges must continue to be an agreeable city in which to live and to that end, the city renovation must continue undiminished. It is no longer intended to increase the number of inhabitants (20,000 people are at present resident in the inner city) but primarily to attend to abandoned buildings and encourage the current varied family composition.
- Trade must remain concentrated in the typical commercial streets, primarily in the main streets leading from the old city gates.
- A concentration of tourism is encouraged in the so-called *Golden Triangle* of Beguinage – Market – Zand Square. Higher quality, overnight tourism is to be stimulated further and the 1996 decision to call a halt to the addition of more hotels must continue to be respected.
- Selective accessibility to the city is to be increased, and the car park and public transport policy is to continue unabated.
- The area around the railway station has been piloted as the nucleus of new city development and as such becomes strategic territory for future large-scale city centre functions and needs at local level, all within close proximity to one another.

The greatest difference between the New Structural Plan and its predecessor of 1972, is the comprehensive examination devoted to the interaction between the historical city and the surrounding region. A new development model has been proposed, its essential aim being to form a coherent structure whereby, increasingly, a central position is allocated to the city and the surrounding region in the relationship between the historical city and her suburbs.

This plan focuses on development. Town planning proposals must contribute to social-economical and cultural development and morphological models must support these developments. New transport concepts must be applied in the balance between space and mobility.

The final phase of the New Structural Plan for Bruges is currently being mapped out. It is expected to be approved in 2000 or 2001.

The extra-district plans

It is important to note that extra-district plans also exist, i.e. the Structural Plan for Flanders and the Structural Plan for the Coast. In the New Structural Plan for Flanders (approved in 1997) Bruges was classified as *a regional and urban entity of the first degree*. In this plan, the historical cities of Bruges, Ghent and Antwerp were declared figureheads for Flemish tourism.

The Province of West Flanders is responsible for the layout of the draft of the Structural Plan for the Coast – which includes Bruges – and is designated as *a living area of great historical and cultural value*, with the result that, in this stuctural plan also, the conservation and restoration of Bruges' heritage has central importance.

As a result the New Structural Plan for Bruges must remain within the framework of the basic assumptions of the above quoted plans and not be contradictory to them.

Management and regeneration action

For 30 years now steps have been taken to control, manage and maintain the heritage and built-up areas by both legal and financial means and by new statutes (e.g. the official protection of heritage laws of 1931 and 1976) (Ministry of the Flemish Community, 1995). The most recent of these statutes are briefly outlined here.

Taxation of longstanding vacancy

Following the Flemish Governmental Decree of 22 December 1995, the city now systematically controls all empty and neglected buildings on its territory. Longstanding vacant, neglected or slum buildings are noted, their owners informed and, where necessary, heavily taxed by the Flemish Regional Government – this is an accumulative tax of a minimum of approximately 7,500 Euro per year. The April 1999 inventory listed 1,108 vacant properties of which 981 will be taxed.

This policy has had an effect and both owners and tenants who are helped wherever possible to solve their problems by a complete series of subsidies from the city, the Province of West Flanders and the Flemish Government. The taxes accumulated by this policy are injected into the Social Impuls Foundation (SIF) whose *raison d'être* is the fight against vacancy and slums.

Listed buildings/monuments
The owners of legally listed monuments – there are around 300 in Bruges – are expected to be 'good house-fathers' to their properties, i.e. to take loving care to keep their buildings in good repair. This has been made easier by maintenance premiums released in 1993 (restoration premiums are also available, see below) (Ministry of the Flemish Community, 1992). In return for the premium, it is compulsory for owners to present an annual technical report to the Flemish Government.

The Monument-Watch – A non-profit making organization.
The provincial and regional governments launched *Monumentenwacht Vlaanderen,* 'Monument-Watch for Flanders', in 1991, to stimulate regular and preventative maintenance of listed buildings. Owners of such properties, as members of the association, are entitled to an annual inspection and detailed report of *their* monument. This usually results in a restoration plan or the carrying out of urgent work. Minimal repair jobs are completed immediately at the time of control by the *monument-watchers*, and whatever the situation, owners are better informed of the physical state of their historical building. Of the approximately 300 listed buildings in Bruges, it has to be admitted that only about 40 owners are members of the Monument-Watch.

The Municipal Department of Historical Monuments and Urban Renewal
There are a thousand and one ways of continuously controlling what is happening in the historic inner city (Beernaert, 1992). It is, after all, the territory of the thirteen members of staff attached to the Department of Historical Monuments and Urban Renewal. Frequently, they are required to act as heritage inspectors and it is they who have direct contact with a building's owners. Permanent care of the historical heritage is, after all, their main function. Negotiations in obstinate cases often take years to sort out, but frequently end, nevertheless, in good results. One example is the eighteenth-century castle of Rooigem and its adjacent park. After years of neglect, restoration has finally been agreed.

As is commonplace, the policy practised is the responsibility of the local City Council, in the shape of the Mayor and Aldermen. In Flanders, this is also true of heritage management. One alderman has specific responsibility for heritage care and urban renewal, but practical tasks and guidance rest on the shoulders of the Department of Historical Monuments and Urban Renewal, which, as mentioned previously, came into being in 1976.

Daily tasks include the control and follow-up of building permission; and each and every request is placed before the Department of Historical Monuments, by which they are able to control the *townscape*. For the year 1998, 596 building permission dossiers for the old city were handled by the department (74 per cent were for domestic buildings, 17 per cent for alterations to commercial edifices, including cafés and restaurants and only 2 per cent were for new buildings). Information regarding town planning (in 1998 this amounted to 180 cases) and town planning certificates (in 1998, 30 cases) were also dealt with by the department.

For each request the following criteria are examined:

- the architectural and cultural-historical value,
- the situation in the townscape,
- and the physical state of the building in question.

The principle still applied – as far as Bruges is concerned – because of the specific character of the city, is that of conservation of its valuable heritage. In no way does this exclude new buildings, indeed this is stimulated at every opportunity where existing architecture is little value.

An extremely important task of the Department of Historical Monuments is advising and following-up restorations whether government or privately owned. Daily visits are made by a building controller and in 1998, 78 violations were recorded.

Much scientific and educational work is also accredited to the Department of Historical Monuments. The annual – and in Flanders enormously popular – Heritage Days, are organized by the department. They play an exceedingly important role in spreading heritage conciousness and especially in increasing the general public's commitment to the architectural heritage. Each building open during a Heritage Day Weekend can usually expect to be visited by several thousand interested people.

Another responsibility of the department's staff is a new examination of the houses in Bruges together with the history of their previous occupants. This is conducted with the help of the non-profit-making organization *Living Archives*. Newly discovered facts often lead to a stronger argument for conservation and sometimes point to a future purpose for the house in question. The Department also lends its knowledge to exhibitions and competitions, thereby helping to raise the profile of the city.

In addition, the City of Bruges has had an Aesthetics Committee since 1904. Its membership includes no architects but is made up of people who have, one way or another, shown particular interest in the history, cultural heritage or architecture of Bruges. The committee is responsible for advising on *aesthetic* alterations to existing buildings or proposed new ones.

City building regulations

Advice grouped together by the Department of Historical Monuments and Urban Renewal and the City Aesthetic Commission is intended to aid the City Council in making their decisions to grant building permission. The council's decision is final.

Bruges exercises an extremely strict policy in regard to City Building Regulations in the whole of its territory including building-plot division and vegetation. In the regulations various articles determine the degree of concern for quality and aestheticism in the field of building:

- Permission granted by the City Council is necessary for: roof renovation (red undulating ceramic rooftiles are obligatory in the historical inner city); alteration of a building's function (introduced to protect city-dwellings); installation of signboards (illuminated advertisements are forbidden in the old city); sunblinds and awnings.
- In a separate ruling, the size and type of advertising boards are carefully controlled – a very necessary regulation in a city as dependent on tourism as Bruges.
- New shop premises are also subject to strict regulations.
- Permission is also needed for alterations to the appearance of a façade (plastering or the removal thereof) painting, cleaning and woodwork renovations. For colour changes, approval has to be gained from the Department of Historical Monuments. The addition of colour on façades is immensely important as the architecture can be emphasized or minimalized or the use of the correct or incorrect colour. The application of mineral paint is requested more and more because of its optimal conservation properties.
- Permission for the restoration and renewal of woodwork is also handled by the Department of Historical Monuments. A great deal of time and energy is given to the advice offered, unfortunately not always with equal success. Joiners appear to have lost many of their traditional craft skills, and often deliver rather banal and standardized work. Recently, however, a few have been found willing to make historically correct doors and window frames. The use of old glass in restoration work is also encouraged. Small details such as these, are extremely important aspects in preserving the general townscape.
- Within the historical inner city, garage doors are refused on façades of less than 12 metres. Garage doors in buildings with a particularly remarkable façade are categorically refused.
- Building materials are also subjected to specific rulings.

The Urban Building-guidance Commission

This Commission, established in February 1996, advises the city council on important urban building alterations. Examination is undertaken of the (urban) development – the bearing surface of a building site and regulations for functional filling-in are sought, note being taken of urban development criteria and suburban regulations. The Committee members include not only the City Mayor and Aldermen responsible for Urbanization, Urban Renewal and Historical Monuments, but also the City Department of Works staff and some external advisers, particularly professors from Flemish Universities.

In the last 3 years, prime locations had to be found for a concert hall with a capacity of 1,200; an exhibition hall; and a 10-screen cinema complex. Two contributing problems were Pandreitje as 'situation' (this is the site of the previous inner city prison), 80 dwellings and an underground car park have now been designated to this area; and the concert hall as 'function', completion of which is foreseen for 2002. An international architectural competition was organized for both projects.

Following definitions set out in the New Structural Plan for Bruges, a draft was prepared by the city of the area around the station; both zonal and functional potential were probed and concrete proposals made.

Financial aspects of restoring and renovating the city

Protection and restoration of the heritage together with general good management is not cheap. Several institutions contribute financially (Esther, 1994) and we will now look at a number of sources in more detail.

The city

The annual restoration premium paid up by the city of Bruges as its share in the restoration of listed privately-owned buildings, for example, was approximately 218,000 Euro in 1998.

On top of this, the city, on its own initiative, subsidizes the restoration of unlisted but architectural historically valuable private property. Since the middle of the nineteenth century the city has been concerned for the conservation and restoration of its heritage (Constandt, 1988). In 1877, therefore, it was decided to grant subsidies to so-called 'artistic repairs' – since this date no fewer than 800 restorations have been supported by the City Council. This subsidy comprises 50 per cent of restoration costs for anything visible from the street, and 30 per cent for the back gable and any worthwhile interior elements. For several years the subsidy has been limited to around 18,600 Euro per building. The city paid out around 471,000 Euro in 1998.

Also important is the 'Subsidy Granted for Functional Improvement of a Dwelling'. This is intended for historically less important buildings with a small land registry tax (around 1,000 Euro). Financial help is given for new roofing, renewal of electricity, bathroom installations and damp proofing. The premium runs to 40 per cent of the work with a maximum of around 4,500 Euro per building. This city subsidy was started in 1979 and the total granted to date is approximately 6,800,000 Euro. This particular subsidy has had an enormous impact on raising living standards, no fewer than 10,500 premiums have been requested since 1979 and as many dwellings have been improved.

A similar subsidy-strategy was worked out to reduce the number of empty properties above businesses. A 40 per cent subsidy is granted for each dwelling installed, with a maximum of around 7,450 Euro or 14,900 Euro per building. The succes rate leaves much to be desired. This phenomenon is present in all European cities and remains difficult to combat.

The city largely pays for the restoration and conservation of her own building heritage, but where listed buildings are concerned, subsidies from a higher authority are also at her disposal (Ministry of the Flemish Community, 1991).

Almost all public property improvements are the city's responsibility. Between 1979 and 1992, for example, no fewer than 14.35 million Euro went into the renewal of streets and squares. The urgently needed canal clean-up operation and modernization of the sewage system between 1973 and 1983, amounted to 37 million Euro (60 per cent being received in subsidies). 2.26 million Euro were invested in the 1980s in large gardens bought for public use and for new cultivation throughout the city. Bruges gives enormous thought to 'green' policy.

The Province of West-Flanders

The Province of West-Flanders, like the city, contributes to the restoration of listed private property and the conservation and restoration of her heritage on Bruges' territory. In recent years, large sums have been spent on the restoration of St Saviour's Cathedral, The Bishop's Palace, The Great Seminary, Tillegem Castle, and the Toll-house.

The Flemish Regional Government

Owners of listed buildings (both private and public) are eligible for restoration subsidies granted by the Flemish Regional Government. Conservation premiums constitute 40 per cent of the work carried out, if less than approximately 14,900 Euro, and 25 per cent of work estimated above approximately 29,800 Euro. The Flemish Regional Government is unilaterally responsible for this conservation subsidy.

The percentage of subsidy granted varies according to the building's nature and is derived from a combination of regional, provincial and local resources (Fig. 2.2). Private buildings are eligible for 40 per cent subsidies (25 per cent from the Flemish Regional Government, 7.5 per cent from the Province and 7.5 per cent from the city). Some private buildings (those belonging to non-profit-making organizations for example) are entitled to

Fig. 2.2 Spinolarei 21. One of the many examples of a listed private building recently restored with the financial help of the city, Province and Region.

an 80 per cent subsidy, but with the proviso that the monument is opened to the general public (50 per cent from the Flemish Regional Government, 15 per cent from the Province and 15 per cent from the city).

Public buildings that are owned by regional administrations receive 60 per cent from the Flemish Regional Government. Churches may receive a 90 per cent subsidy for restoration (60 per cent from the Flemish Regional Government and 30 per cent from the Province). The city may also receive an 80 per cent subsidy for the restoration of the buildings it owns (60 per cent from the Flemish Regional Government and 20 per cent from the Province). School buildings have a special ruling. Tax deductible methods have recently been introduced to stimulate contributions from the private sector.

New uses for historical buildings

A historical building's present function or re-use is closely monitored due to the obligatory licence. An 'inherited' use, i.e. a dwelling, generally causes no threat to the architectural heritage, but it is always advisable to monitor the situation. Interesting building elements disappear almost daily from the Bruges scene, despite the enormous effort made and control imposed by the Department of Monuments.

In so-called 'heavy' functions, i.e. offices or hotels, problems are mostly unavoidable. For each application, it is necessary to find the right balance between the wishes of the Department of Monument's heritage experts and the Flemish Governmental Offices on the one hand, and the wishes of the person commissioning the work on the other – his or her main interest being the most economical means of completing the project. The procedure is reminiscent of tight-rope walking until an acceptable compromise can be agreed. However, there are, happily, some good examples of re-use to be found in Bruges.

Vacant historical buildings

Acute problems with large vacant properties are minimal in Bruges, but there are a few exceptions. These last are closely monitored by the City Council (see above). Re-use and/or mixed purposes for church buildings are among the problems which are likely to arise in the near future. The attention and close monitoring necessary are already becoming apparent and the Church Authorities themselves are looking into the problem and trying to find solutions. It is obvious that the city also has a duty in this respect. Re-use of the religious architectural heritage is a problem in several European countries and generally concerns buildings of great architectural and cultural-historical value. Is this not an ideal area for the development of collective ideas?

Council dwellings

The City of Bruges had always striven to maintain the number of council dwellings in the historical inner city. Recognized social building firms have realized no less than 800 homes since the early 1970s, many on strategic sites; e.g. 120 appartments built between two shopping streets in the heart of the city centre. A small number of houses owned by the City of Bruges are rented out, usually at a social rate. Additionally, there is the role of the OCMW (Public Commission for Social Welfare): from the time of the French Revolution, they have managed 48 *godshuis* complexes in the inner city. These are historical residential entities around an inner court or garden and are to be found throughout the inner city, providing very comfortable cheap accomodation for 300 pensioners. They have been systematically restored and renovated over the last decade of the twentieth century.

Environmental management

Traffic is the bane of every historical town. Car possession has increased enormously since the 1960s and 1970s and drivers are intolerant of the smallest hindrance to their mobility (Keppler, 1987).

Traffic

In the Structural Master Plan of 1972, Bruges had already developed a new vision for coping with traffic. This was in the shape of three traffic schemes, i.e. saturation, optimal and transitional. The optimal scheme foresaw car traffic being greatly reduced in the historical city, but it was soon recognized that this would be unrealistic and it continues to be so. The Structural Plan opted for the so-called transitional scheme in 1973, a loop-system conducting traffic in a diagonal direction thereby creating a traffic reduced pedestrian area and a ring of graded parking facilities. A new circulation plan was implemented in 1978, realizing the most important elements of the transitional scheme. This traffic model worked well up to the beginning of the 1990s.

Alterations to transport circulation relying mainly on one-way systems, was coupled with intensive renovation of streets and squares. This was certainly a change for the better, not only invigorating the ancient city's character, but also making life more agreeable for city dwellers. Renovated squares quickly became new meeting places for the inhabitants and places for city functions; they often were an incentive for the restoration of surrounding buildings. The transport circulation plan was ahead of its time and became an example to other cities both at home and abroad.

The period 1978–92 saw an important growth in mobility hindrance. The number of cars rose by 62 per cent and traffic volume in some streets by as much as 30 per cent. This was all due to an increase in the number of people living alone, a growth in one-day tourism, the strengthening of the city centre's functions and an increase in the service sector. On the other hand, public transport saw an enormous drop in usage (45 per cent between 1975 and 1990), long-term parking met with little resistance (6,000 cars repeatedly parked long-term in residential streets) and guidance for bicycle use was inadequate. Through-traffic was reduced by the circulation plan of 1978 but this did nothing to resolve the intensity of car transport (10,000 traffic movements per peak hour).

The transport circulation plan was again altered in 1992, this time to cut off the old Market Square to through-traffic and to facilitate and encourage public transport and bicycles. A 'zone 30' was introduced for the entire inner city; this entails a maximum speed of 30 km/hour for all motorized vehicles. In the meantime, Bruges has grown into a true bicycle town. Thousands of bicycles are allowed into squares and streets where motorists are prohibited by traffic barriers. In truly pedestrian zones, bicycles are also banned.

Thirty-five new buses were bought by the public transport company *De Lijn*. These are extremely manoeuvrable and better suited to the ancient city. They are also smaller, more environmentally friendly and have lower floors for easier access. Special bus-lanes give priority to buses and at certain places they activate traffic lights which give them preferential treatment when changing direction. Since 1992 the buses run more frequently. In September a free monthly pass is handed out to school children in the hope that they will continue to use public transport for the rest of the school year. The new rules seem to have passed their first assessment well.

Parking policy

More than one-quarter of inner city car owners have the use of a garage (in 1996 there were 7,782 private cars registered). The rest are parked in streets or local open-air car parks. The city itself built fifteen small-scale local garages – an elegant solution for its resident's parking problems, but this is rather inadequate as there are only places for 763 cars. Over one hundred local garages were built by the private sector, with a capacity for 2,273 cars. The biggest problem at the moment is signalled in the Steenstraat and the Burg, where the commercial centre is concentrated. A recent analysis rates the shortfall to be at least 1,300 private parking places. Residential parking is undergoing a trial period in a few streets. A number of parking places are reserved for local inhabitants, for which a resident's card is needed.

Some activities, particularly schooling and shopping, are attracted by private transport. Solutions are permanently being sought for the acute shortage of parking places. Four underground car parks have been built in the last 20 years, with a total capacity of 2,032 places (Zilverpand for 420 cars; t'Zand for 1,200; Beguinage for 200; and Biekorf for 212) (Fig. 2.3). An open-air garage at the station has room for 1,600 cars. The bus company *De Lijn* provides a cheap and efficient connection from the car park to the city centre. At the Katelijnepoort, a coach park has been installed for 90 tourist coaches. Parking meters in many streets and squares favour the short-term visitor.

The threat to the historical heritage by motorized vehicles is now less acute due to a reduction in petrol fumes and building vibration. Atmospheric pollution is a general problem not specific to Bruges, but the city does have the advantage of not having any strongly polluting industry in its immediate vicinity.

Tourism and heritage management

More than 3 million people visit Bruges each year; approximately 1 million staying overnight, the rest only for the day (Maes and Esther, 1999).

Fig. 2.3 t'Zand – underneath this big square an underground garage offers space for more than 1,200 cars.

Bruges' attraction is due not only to the particular character of its heritage and medieval street pattern but also to the special atmosphere created by many small-scale multifunctional activities. The city is not a 'theme park' with a varnish of culture on the top but rather a city with a lively and authentic culture waiting to be discovered by the interested visitor, and at the same time providing a certain quality of life for its inhabitants. The inner city's character must be preserved at all costs.

Mass tourism must be handled with the necessary caution and mistrust, otherwise problems are bound to occur that could jeopardize the city's specific attraction. First and foremost, there is the possibility of direct damage to the historical heritage itself, but because visitors are spread out, due to a large choice of both museums and monuments, the toleration-ceiling has not yet been reached.

The most successful tourist locations are visited yearly by between 750,000 and 900,000 people. Such places, for example, the Beguinage (Fig. 2.4) and the Church of Our Lady (Fig. 2.5), are easily able to cope with these figures due to the size of the sites and buildings. Nevertheless, it is still necessary to protect the character and function of these two sites as well as the interests of the Beguinage's residents and the church's users. Conflict has already appeared at the Church of Our Lady; this is both an acting parish church and a religious cult building, but it has also had to contend with thousands of tourists on their way to visit the graves of Mary of Burgundy and Charles the Bold, and to see Michaelangelo's Madonna. For these reasons, the Church Authorities decided on the following measures: tourists are refused entrance during religions services; guides' explanations are not allowed within the church and in order to promote more respect for the

Fig. 2.4 The Beguinage was founded in the thirteenth century. This beautiful enclosure is now occupied by Benedictine nuns – the small houses are rented out to ladies living alone. Each house has been carefully restored by the City of Bruges. (Photograph: Jean Godecharle.)

church building, recorded religious music is played during tourist 'peak hours'. The Belfry, St Saviour's Cathedral and The Holy Blood Basilica have 500,000 to 600,000 visitors annually. Only 70 visitors at a time are allowed on the Belfry's tower.

The most popular museums have annual visitor totals of 150,000–200,000. These are the Groeninge (with the most important collection of old paintings), Memling (with the medieval hospital of St John), the City Hall's gothic hall (with its nineteenth-century wall paintings) and the Palace of Gruuthuse (mostly applied art). Other fascinating and interesting monuments and museums achieve only approximately 60,000 annual tourists; but this results in a more intimate and relatively higher quality visit. Up to the present time, no damage has been recorded due to the number of visitors.

Small, fifteenth-century monuments, such as the Bladelin Court in the Naaldenstraat and the Jerusalem Chapel in the Peperstraat, can cope with only very few visitors. Publicity is consciously minimal, enabling them to be discovered by the truly interested and usually individual visitors.

It is often difficult to balance the tourists' expectations and the residents' needs and habits. The social acceptability of tourism is difficult to judge. Are the residents happy with the situation? It is only occasionally that things have gone too far. Ten years ago a public protest movement raised its head, the 'S.O.S. for a bearable living style for Bruges' residents'. This movement was especially concerned with the increase in tourism; the distress call was taken to heart by the City Council and several compulsory laws were introduced, e.g. a stop to further hotel expansion. Furthermore a tourist-concentration model was launched in the so-called 'Golden

Fig. 2.5 Tower of the Church of Our Lady. The tower's restoration is planned in the very near future.

Triangle', thereby discouraging a greater amount of tourist activity from the rest of the city. Greater priority was also given to cultural tourism.

Another very real danger caused by the tourist explosion, is the possible repression of dwellings. Certain commercial concerns follow in the wake of the tourist, i.e. lace and chocolate shops (both specific to Bruges), souvenir shops, cafes, restaurants and hotels. 'Strong' functions force up property prices and push out the everyday consumer shops, i.e. butcher, baker and grocer. However, the City Council monitors the situation by imposing the necessary building licence on buildings requiring a change of function. Even in the 'Golden Triangle' tourist-linked functions are not permitted in some streets.

At the present time there are 106 hotels in the historic city, with capacity for 6,075 guests (an increase of 15 per cent since 1981). Youth hostels, campsites and bed and breakfast accomodation in the inner city and suburbs, cater for another 2,500 visitors. The rate of hotel occupation in 1995 was 55 per cent on a yearly basis, and appears to be increasing.

Whilst the tourist industry in Bruges is only responsible for 8 per cent of all jobs, the economic value of tourism obviously must not be overestimated.

Sustainability

The total process of urban renewal and monument care must continue to be supported by both a structure and acceptable future planning.

The New Suburban Structural Plan

The future seems assured for Bruges. The Department of Historical Monuments – by Belgian standards, of quite generous proportions – is prepared to keep a close eye on changes and alterations in the inner city. The new Suburban Structural Plan as a possible handbook for the future, is expected to be authorized in the near future. This, like the Structural Plan of 1972, will need adaptation from time to time as certain needs alter or unexpected changes occur. Eventual adaptation will be the responsibility of a City Urban Renewal Guiding Commission similar to the one already functioning in the borough legislative framework of 1995–2000. It is comforting to know that there are enough existing legal means which can eventually be amended and that finance, to a greater or lesser degree, will be made available.

An inventory of the historical heritage

A systematic inventory of Bruges' historical building heritage was begun in 1997 by the Ministry of the Flemish Community. Four volumes are devoted to Bruges in the series *Building through the Centuries.* The series is expected to be ready in 2002. The inventory is seen as leading to new legal protection for buildings, thereby assuring their future. The City of Bruges is working intensively with this project by delegating two staff members to work on the inventory and by lending the rich knowledge and enormous number of documents of the Department of Historical Monuments.

The city would like to see more buildings on the protected list. This may be assisted by the fact that the Flemish Regional Government's restoration budget for legally listed buildings has trebled since its commencement and the success of the Heritage Days in 1989, and is expected to continue to increase.

Cultural Capital of Europe in 2002

Together with Spain's Salamanca, Bruges in 2002 will be the Cultural Capital of Europe. This title is far from unimportant, and not only for the cultural aspect or future fame of Bruges. The 2002 project is an important evolutionary step as rapid development is made possible. Ambitions promoted by 2002 are directed also to the architectural and cultural heritage and efforts made for the 2002 project must have a lasting effect on cultural direction, monument care, public property, (new) architecture, and tourism. Restoration is planned among others, for the spire of the Church of Our Lady; for seven houses in the Beguinage; important maintenance to the Belfry Tower; the façades of the Griffie (Fig. 2.6) and The Palace of the Liberty of Bruges on the Burg Square; the wall paintings in the City Hall's gothic hall; the Academy of Fine Arts; the Music Conservatory; the Theatre and the former Jesuit College. A figure of 25 million Euro has been estimated for all this work; extra finance is also possible. The building of a new concert hall has already started and it is also expected to open in 2002.

It is very important that the architectural heritage and museum collections are opened up by means of specialized guiding, good brochures, responsible publications and audio-visual material. This will help to realize the aim of spreading a more correct knowledge of the heritage, even eventually enabling it to be beloved. Interdisciplinary co-operation between diverse heritage specialists seems to be an obvious necessity. For the tourist,

Fig. 2.6 Burg, the Griffie. Originally a brightly coloured building of the sixteenth century and one of very few Renaissance-style buildings in Bruges; the Griffie is due for restoration before 2002.

reception, help and guiding are extremely important elements in cultural activities; and, we must never forget, for the residents also.

The Historical Towns Association for the Euroregion Kent – Nord Pas de Calais – West Flanders – Hainaut (VHSE in Dutch)

This association of which Bruges was a founder member in 1995, now has 21 members. Its chief aims are first to set up a network for co-operation and exploration of the experience gained by the historical towns of Europe in the aspects of environmental management and conservation; and second, to encourage economic activity compatible with quality of life, and the preservation of Europe's urban heritage. The *VHSE* intends to be a valid partner in discussions with, and a means for addressing, the European Union (EU). The EU has exerted itself greatly in the past on behalf of large cities, now attention is being sought for towns of a lot less than 100,000 inhabitants. Is this an opportunity waiting to be seized?

Candidature of the UNESCO – World Heritage Convention concerning Cultural and Natural World Heritage

The Convention's ratification charter was only deposited with the Director General by the Belgian representative to UNESCO on 24 July 1996. It is therefore only now possible for Belgium to catch up with other countries in regard to placing their most important heritage on UNESCO's World Heritage List.

The Flemish Government was responsible for initiating an application to UNESCO for the Beguinages. Furthermore, Bruges' Beguinage was recognized as a World Heritage Site on 2 December 1998, as an 'expression of a

Fig. 2.7 Panorama of the very heart of the city.

unique tradition going back to the Middle Ages, and of an extremely valuable architectural character'. In the meantime an application has been made to include the belfries as symbols of Flemish Urban Autonomy. Bruges' monumental Belfry, is one of the most typical.

Finally, on 1 July 1999, a large and comprehensive file was presented, requesting the complete historical inner city of Bruges (Fig. 2.7) to be recognized as a World Heritage Site. A result is expected before 2002; an eventual recognition will bring future guarantees for Bruges' heritage.

Conclusions

Although the above activities, plans and results give rise to certain positive conclusions, a watchful and critical eye must always be open.

During the last 30 years, the City of Bruges has succesfully brought about a movement of awareness, first on the level of their own policy and later to the general public, architects, building contractors, and (to a lesser degree) project developers. The challenge in the future is to find a harmonious way for growth in this urban fabric, calling upon all possible partners to work together. Maintenance carried out in time, cautious repairs of important architecturally historical buildings (initiated by or with help from the government) and new building can only contribute further to the perpetuation of Bruges' valuable and unique character.

A new restoration movement backed up by urban development support, appeared in the 1970s, its emphasis came to be the restoration of ordinary homes. This evolution continues to this day. It is worth noting that in the 1980s, the private sector was responsible for many restorations, the local government generally being allotted the role of supervisor. At this time, however, the City Council gave much thought to making public property more attractive; this in turn, led to improving the value and worth of

Fig. 2.8 More than 3 million tourists visit Bruges each year.

buildings in general. The trend today is much less obvious, the city's contribution has increased and large monuments are next on the waiting-list.

The choice between conservation or new building remains a sensitive issue. Unimportant buildings can simply be knocked down and replaced by new. Fortunately, Bruges has several, though scarce, examples of successful new architecture. It is of prime importance that new architecture reflects the culture of today.

What is meant by a valuable heritage, has undergone striking change in the last 30 years. Buildings formally classed as architectually unimportant are now appreciated and protected; demolition is now the exception rather than the rule.

Heritage conservation seems to be well looked after by urban development guidance and regulations. Finance must continue to flow for this most important aspect of European society, and it is essential that approved urban development policy is respected by subsequent political coalitions. Political will is a primordial necessity to any realization. The problems of transport and mass tourism must be watched extremely carefully (Fig. 2.8). When necessary, a turn-about must be made possible.

For all feasible aspects, it should go without saying that the key word is quality. This would appear to be the message for the twenty-first century.

References

Beernaert, B. (1992) *Historic Preservation in Bruges: Continuous Planning in operation*, in Dutt, A.K. and Costa, F.J., *Perspectives on Planning and Urban Development in Belgium*, Kluwer Academic Publishers, pp. 147–91.

City of Brugge (1976): *Structuurplan voor de binnenstad*, Brugge.

Constandt, L. (ed.) (1988): *Stenen Herleven. 111 jaar kunstige herstellingen in Brugge, 1877–1988*, Marc Van de Wiele.

Miloš Drdáchý

Telč, Czech Republic

Introduction

Telč is a Moravian medieval town, founded in the thirteenth century as a fortified place with two castle-fortresses, town wall and lakes around (Fig. 3.1). The lower castle, originally Gothic, was rebuilt in the sixteenth century into a Renaissance chateau with rich, well-preserved interiors. In the seventeenth and eighteenth centuries the city was enriched with Baroque churches and sculptures, mostly built by Jesuits. The Gothic period is apparent in St Jacob's Church with its Gothic frescoes and the Church of Mary's Assumption in the Old Town. The Church of the Holy Spirit with a Roman tower dates back to the thirteenth century. A unique, well-preserved complex of Renaissance and Baroque houses on the main square form one of the most important Czech architectural monuments. In 1992 Telč was designated as a World Heritage City by UNESCO.

The city of Telč and its surroundings have been the subject of Technical Assistance and Follow-up Programmes supported by the Cultural Heritage Department of the Council of Europe since 1993. In Telč this programme studied in particular the possibility of applying experiences with the so-called safeguarding and enhancement plans prepared for French historic centres.

However, the conditions in the Czech Republic call for an individual approach that might be of interest for other small historic cities and their micro-regions. Telč itself is a World Heritage City with about 6,000 inhabitants and less than 100,000 cultural visitors coming from other parts of the country or the world. This fact has to be reflected in the strategy of the city development, as well as in the local resources affordable for safeguarding the cultural heritage.

The programme to create a strategy on social and economy development in Telč is based on four basic items:

Fig. 3.1 Roofscape view of Telč. (Photograph: Council of Europe.)

- the designation of the city as a World Heritage Site;
- a natural centre of the micro-region;
- a city in which the unique character must be protected for future generations; and,
- as a high quality place for the life and work of its citizens.

To implement this strategy several tools are prepared and introduced into the management practice (Drdácký *et al.*, 1998). They will enable decision-making built on a better knowledge of local cultural heritage and on more detailed data on community needs and potential. However, it must be emphasized that the process of introducing new management tools is still in the process of establishment.

The policy and planning framework

Preservation and enhancement of historic areas

Territorial protection of monuments within the Czech Republic has been ensured by law since the 1950s and the preservation of urban or landscape complexes is increasingly emphasized. There are 40 town heritage reserves, 61 village heritage reserves and 10 archaeological monument reserves. Cores of historic cities are further protected by 209 town heritage zones and 164 village heritage zones. Cultural heritage landscape areas are preserved in 17 landscape heritage zones.

The Czech Government adopted the Granada and Malta Conventions in 1998. The Ministry of Culture recently made an analysis of these Conventions with respect to the legislative situation in the country (Matoušková, 1998; Pickard, 2001). Some criticism may be levied at the present system in light of the Conventions, noting, at the same time, the progress that has been achieved in developing a documentation system and also the progress made in developing integrated urban planning methodology for conservation (see below). These actions are being developed to solve most of the existing deficiencies concerning the articles of the Conventions. Actions being developed for Telč may assist the process of improving the mechanisms for all Czech historic centres.

Among the values of assessment on monuments defined in article 1 of the Granada Convention (definition of the architectural heritage), 'archaeological' and 'social' importance is not considered by the Czech Law on State Monument Care (No. 20/1987). Moreover, the Convention applies homogeneity, coherence and character of architectural ensembles or sites, which the Law also does not accommodate. This may create problems with the topographic determination and the urban character value recognition of ensembles of cultural heritage interest.

In respect to article 4 (supervision and authorization measures), the necessary checking procedures exist. Authorization certificates are issued for persons carrying out restoration works and archaeological works, but this authorization does not concern the design and construction processes. Unfortunately, the present law and especially its sanction provisions are not capable of preventing distortion, degradation or destruction of the protected cultural heritage. The responsible authorities evaluate intervention projects on listed or listing-proposed buildings, as well as changes in their neighbourhood, but only from the point of view of the individual building.

The character of an architectural ensemble or a site is not defined as a cultural heritage characteristic in the law and hence not considered. The impact of interventions on heritage might be assessed using corresponding laws, No. 244/1992 on Environmental Impact Assessment and No. 50/1976 on Land Use Planning and Building Code. However, their criteria are quite far from the purposes of heritage preservation and, moreover, the competent authorities are not sufficiently engaged in cultural heritage matters. Both the Heritage Law and the Building Code give rights to the state to impose necessary works on the monument's owner, but the natural owner cannot be substituted by the state in the case of such works not being carried out. An expropriation is legally supported by both codes but in practice not used.

Public institutions for the maintenance or restoration of monuments do not exist in the Czech Republic, therefore the financial support of public institutions, according to article 6, concerns only the Heritage Institutes which are professional bodies of the state protection service. There are some opportunities for tax reductions but these are minor and negligible. In any case, none of the existing measures encourage opportune and regular maintenance and hence the preservation of authenticity of historic objects.

The Building Code provides the powers to enforce the demolition of a new building which has been built without permission but there is no way to ensure the restitution of the original state of a listed monument (article 9). Whilst article 10 of the Granada Convention emphasizes the need for integrated conservation approaches there is no legislative means at present within the Czech Republic for enabling 'the protection of the architectural heritage as one of fundamental goals of town and country'. Actual laws do not enable the definition and legal proclamation of guidelines for safeguarding and care in territories of cultural value or in protection zones within the planning process. This is because protected areas are designated by the government (or by the Ministry of Culture) and the legal regime does not permit the imposition of duties above its framework (such as in relation to territorial planning). Furthermore, urban planners are not obliged to deal with such territories and they are mostly not properly trained in heritage protection matters. Thus the integration of urban planning and cultural heritage protection is very weak and not legally required.

The adoption of the Malta Convention will require similar legislative measures because an integrated conservation programme is also lacking in this field. Here the threat of the destruction of an important archaeological heritage is more dangerous, as has been unfortunately proven in recent years in the form of uncontrolled developmental activities.

Documentation system for small historic cities

Despite the problems mentioned above, a documentation system for small historic cities such as Telč has been developed as a practical tool for:

- preparation of their strategic plans,
- preparation of land use plans including safeguarding and enhancement plans,
- management of their sustainable development,
- documentation of their historic values and management of monuments,
- inventory of the cultural heritage in a territory,
- preparation of financial and regeneration plans for the built heritage,
- heritage impact analysis of investment in a territory,
- elaboration of information issues for citizens and visitors,
- possible connection to public information systems (Internet),
- research purposes.

The system has been designed as a set of special databases attached to a true geographic information system (GIS). For small cities and villages, a cadastre (land register) digital map is used as a reference base, which enables the recording of relevant data – alpha-numeric, textual or graphical – to individual lots or objects projected on the map. It is supposed that the vectorization of cadastre maps all over the country will be completed by the year 2005 and then digital maps will be available for all cities and villages in the Czech Republic.

The system has been developed in accordance with the Recommendation of the Committee of Ministers of the Council of Europe No. R (95) 3 on 'Co-ordination of documentation methods and systems for historic buildings and monuments of architectural heritage' (1995). Similarly, international standards for archaeological monuments and sites, as well as mobile objects are being applied (Thornes and Bold, 1998).

Existing national standards and recommendations for territorial identification and cultural heritage inventory or documentation have also been incorporated into the system. In order to facilitate the transfer of data already recorded into the developed documentation system, compatibility with ministerial or regional databases for protected monuments has been maintained.

Contents and description of databases
Data for the documentation system of historic cities are collected across a broad spectrum, which enables the construction of targeted databases and outputs. Those databases are grouped into thematic clusters for the sake of brevity. All available pieces of information on monuments, objects of historic or cultural interest and their context links in a territory of interest, are recorded. An important part of the system is devoted to economic data that can be utilized for planning, as well as for heritage impact analyses or potential evaluation in historic cities. Their interaction with environmental issues influencing conservation or maintenance costs is also taken into account. The system provides users with references to documents kept in different archives.

The main groups of collected data and a brief description of their contents are presented further according to eight themes. A detailed structure of actual records is dependent on the type of unit under consideration and it is modified automatically at the moment of its insertion.

1 *The unit definition* contains blocks of basic identification data, related unit references and identification of the data provider. The types of units selected are elements, objects, functional ensembles, historic ensembles, public spaces and territories.
   - *Element* – defined as a part of an object or a parcel of land. It might be a part of a building, spatially defined (wing, tower, gate, etc.), a structure (vault, pavement, door, etc.), art-work permanently built into the object (fresco, stained-glass window, etc.), or a vegetation or landscape arrangement of the lot (balk, terrace, pavement, etc.).
   - *Object* – considered to be a spatially enclosed or functionally independent building (house, barn, fencing wall, etc.), a parcel of land, a spatially independent work of art (statue, etc.), cult (cross, Crucifixion column, tomb, etc.), technical or other functional quality (milestone, well, etc.).
   - *Functional ensemble* – groups objects and lots spatially or technically joined and intentionally linked to fulfil a common function (housed with auxiliary buildings, fences and garden; farm; water mill; sculptural complex; fountain with connected water work, etc.). A functional ensemble is typically identified by one land register number.
   - *Historic ensemble* – contains groupings of generally independent objects and lots, each having an individual function, joined together by historical, cultural or geographic relations (road of the Crucifixion

with Calvary in a landscape, historic core of a settlement, subway network, housing estate, etc.).

- *Public spaces* – includes squares, streets, town parks, embankments, lakes and rivers, etc.
- *Territories* – denotes parts of the Earth surface defined by laws, approved in urban planning documents or other codes and composed of land elements (Town Heritage Reserves, Town Heritage Zones, Landscape Heritage Zones, urban zones, etc.).

2 *The unit location* is described by blocks of administrative determination, historic and geographical determination, address data, cartographic data, cadastre data and land unit price data.

3 *The property and legal data* contain blocks of ownership data, user data, historical review of owners, limits of use rights.

4 *Archive references* are comprised of monument protection documentation, graphic documentation, text documentation, ethnology documentation, environmental documentation, and monitoring reports.

5 *The technical description and condition* include blocks of basic data, technical description of elements, physical condition of elements, utility equipment and supply network, arrangement of the lot.

6 *The unit function* embodies blocks of data on the contemporary function of the unit and its original functional use.

7 *Historical data* contain group blocks on architectural history of the unit, relation to historical events and personalities, and the unit's own chronicle.

8 *Cultural heritage evaluation and protection measures* appraise the relevant features according to individual elements, stories, as well as the entire unit. This group serves not only to describe the listed monuments but is used, moreover, for recording cultural heritage aspects of other objects and sites. In this respect, for example, a methodology developed by Zuzana and Jiří Syrový (1997) is used for rural architecture. An illustration of such a survey is presented for a part of a Telč suburb (Fig. 3.2).

Structure of the system

The basic structural chart of the documentation system is shown in the legend to Fig. 3.3. It is composed of three layers of data structures and programs. Data structures contain the main database and files of digital maps. Raster maps are stored as bitmaps (*bmp*), vector data use a special format. Exporting and importing of vector maps is possible in formats *igs* or *dxf*. The map server is used to facilitate access to application programs on data. Application programs ensure the map management, the GIS function and access to databases. The first package serves for creation, updating and presentation of vector or raster maps. The possibility of users creating and actualizing data on a place is particularly stressed because the system is aimed primarily for management and decision-making in historic cities and must be easily up-dated. GIS programmes enable a search in databases in relation to map entities and vice versa, as well as graphical interpretation of the found data. The system is also intended to be

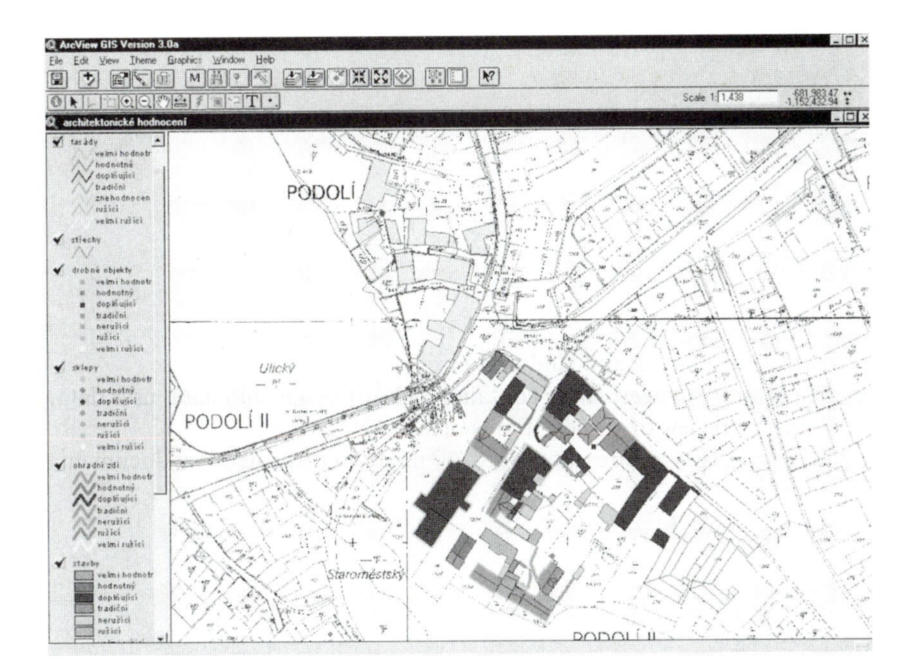

Fig. 3.2 Building archaeology survey record in the Telč Documentation System.

accessible via the Internet for public dissemination of 'open' data and thus facilitating access to remote databases.

Recording of cultural heritage in a micro-region

One of the accompanying actions to the follow-up programme in Telč supported a GIS based recording of a complex natural and built heritage over a broader territory and in context with surrounding historic settlements. This pilot project resulted in a rich database of vectors of collected characteristic data from more than 6,000 land elements in the selected area of 230 sq. km. and prepared a good foundation for continuing the development of appropriate tools for the management of the cultural heritage landscape. The term micro-region is used here for a territory with significant interior, historic, cultural, functional, social and spatial links, the management of which is usually ensured by the historic town under consideration. The above mentioned documentation system is adopted in the micro-region too.

Integration of land-use plans with heritage safeguarding and enhancement plans

Further progress for improving the system of heritage protection for historic centres has been undertaken through the assistance of the Council of Europe's expert consultants in the Follow-up Programme for Telč. This examined the possibility of using the system of French 'safeguarding and enhancement plans' for historic centres as a method of integrating heritage protection into the process of land-use planning for Telč. In developing these ideas expert consultants agreed that it was necessary to find an

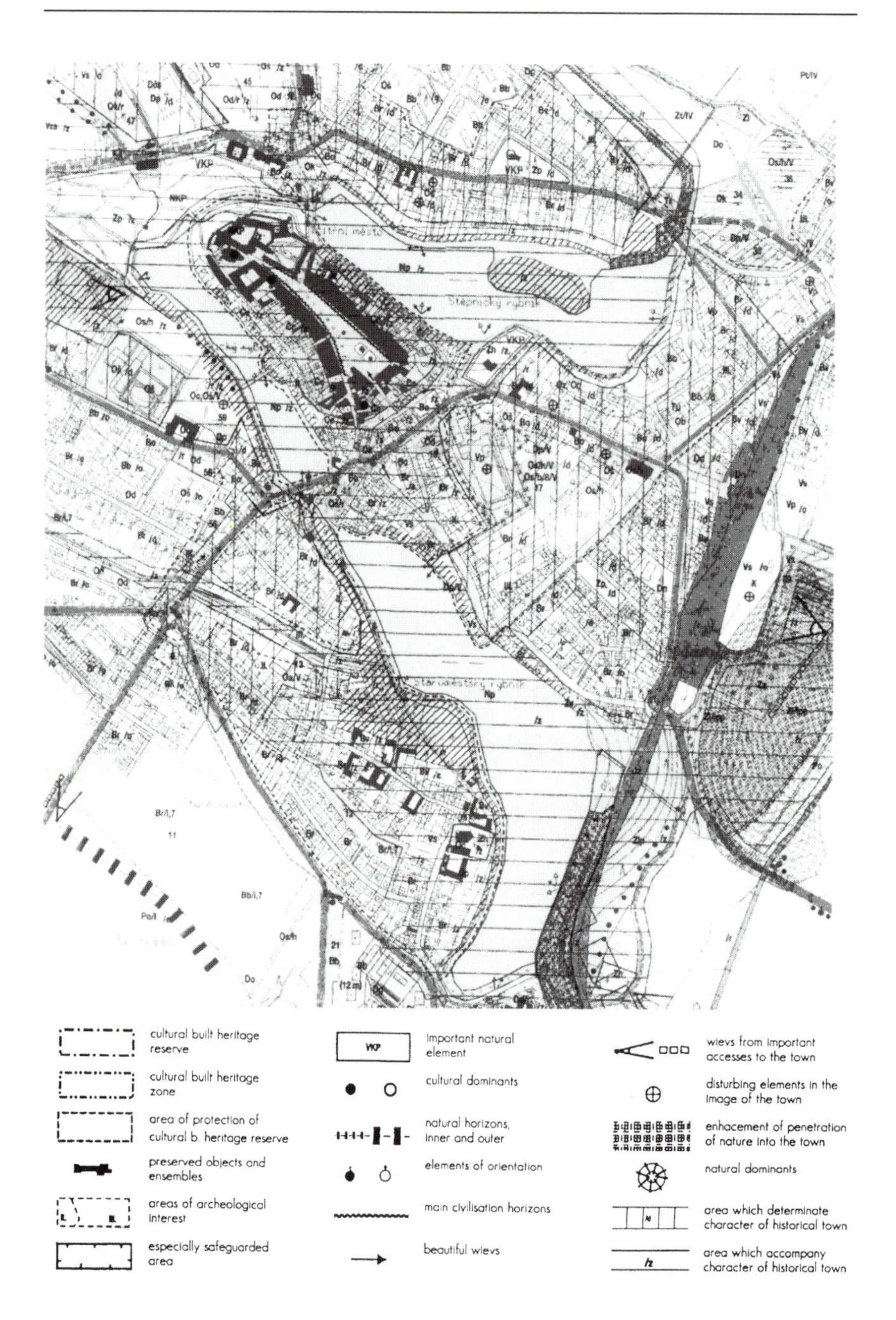

Fig. 3.3 Drawing documenting cultural heritage sustainability requirements in recent Telč Land Use Plan.

approach suitable to the political and historical conditions inside the Czech Republic. In this context, a structure of basic urban planning tools for safeguarding the built heritage in historic centres has been devised. The qualities of the urban structure of a historical town are considered as the basic source of information which can be used to develop policy approaches for rehabilitation and documenting the values of the urban heritage that should be saved in the future. The integrated planning process is being developed in this way.

Analysis of cultural and historical heritage
It is a very slow process to achieve a detailed knowledge of the cultural built heritage that would be sufficient for analysis. It requires the execution of building archaeology surveys, which is very expensive and takes a long time. It is impossible to reach such knowledge within a short time in many towns.

On the other hand the most precious historic towns and villages in the Czech Republic are currently protected as cultural built heritage reserves (MPR) and zones (MPZ) which are proclaimed on the basis of an elementary cultural heritage surveys of towns and objects. So, if a plan to safeguard and enhance any already protected territory is going to be elaborated, it is possible to profit from the existing analysis. However, these analyses usually do not reach the level necessary for determining concrete measures for each object of heritage value. If the plan concerns other territories, (without the existing analysis), it seems rational to elaborate a preliminary simple survey, useful for the categorization of objects according to a degree of protection, but having in mind that it is not sufficient to determine concrete arrangements for each object. Detailed investigations can then be requested for selected categories that will form the basis for the imposition of regulations concerning the issuance of construction permission. The tools to enhance the cultural heritage should be included in both the plan and the regulations.

A historical and artistic analysis of the cultural heritage does not provide all the necessary information that would be required for a plan of safeguarding and enhancement. The social and economic potential of historic buildings and the environment of the area also need to be considered. An analysis of historical urban development and the morphology of the city landscape, including its connections with the surrounding natural countryside, are also enormously important.

Program of enhancement of historical monuments and towns
This phase is considered as the most important part of the whole process and in the national terminology could be called the strategic plan (ÚHZ) for the safeguarding and enhancement of a historical town.

The essential aspect is to find ways of allowing new development that takes advantage of the historical potential but at the same time protects the historical substance. It is believed that such an urban planning concept is the basis of the success of the safeguarding plans in France (*Plan de sauvegarde et mise en valeur*). These types of plans are partially covered by Czech public by-laws on urban cultural heritage reserves and zones (MPR and MPZ). It is worth mentioning that this urban planning quality of French plans is not perceptible in Czech examples of methodology of safeguarding mechanisms which are developed by the relevant conservation authorities and do not integrate with the territorial planning system. The dominant opinion of Czech heritage specialists is that the safeguarding plan should only deal with conservation issues.

It is therefore not easy to find and formulate appropriate urban planning tools, particularly in small towns or in towns with a small historical core. Some towns could profit from the experiences of regeneration programmes which have contained social and economic elements from elsewhere and where public interests have been supported (particularly through programmes of financial interventions).

After many discussions with experts and experiences with the application of the safeguarding plans, the preferred approach in the Czech Republic is to develop an integrated planning system in two steps. The first would be through a 'General plan of cultural heritage sustainability' (the *General Plan*) and the second, a ' Plan for the safeguarding of cultural heritage' (the *Safeguarding and Enhancement Plan*). The tools to enhance the cultural heritage should be included in both documents, although this is not stressed in their titles.

General plan of cultural heritage sustainability

The basic objectives of the General Plan are:

- to save and to conserve historical cultural values in an area before or during the process of preparation of urban plans, a land-use plan or a regulation plan;
- to keep the context and the continuity of the natural, landscape and urban elements in a territory;
- to propose extensive rules to accomplish the preservation of an historical part of town as an exceptional ensemble in its unity and coherence and to develop a town in its totality;
- to constitute an urban planning document which would become a compulsory part of the statutory town plans;
- to identify and to record elements of cultural heritage value, including those elements that are not subject to specific legal protection.

To adequately cover the monuments' preservation needs, both the urban planning considerations and the artistic/cultural activity should be the basic principles for the preparation of the General Plan. Discussions among Czech specialists have shown a tendency not to fulfil this principle, but any other approach would fail in advance. Just a pure conservation approach would not be sufficient – the functions of the buildings and the area also need to be considered. Maintaining the vitality of the area together with the real substance of historical towns can be made possible only by the suitable use of heritage assets.

The detailed contents of the General Plan approach are described elsewhere (Drdácký, 1999), but a basic outline can be presented here. The General Plan consists of a 'graphical' part and a 'written policy statement'.

The suitable scale for the 'graphical part' is between 1/5,000 and 1/10,000. The following information is presented:

- (typical) survey and analysis drawings;
- synthetic drawing of all historical and natural qualities in a territory, divided in accordance with the categories mentioned in the chapter about methodology;
- space and functional regulations, especially the regulation of elevations;
- indication of all important elements of 'image' of a site, especially cultural dominant features, inner and outer landscape horizons (skylines), alleys, memorable trees, important views, disturbing objects, natural dominant features and outstanding terrain forms, areas which must not be built up (from the point of view of conservation), penetration corridors of a city and landscape, limits of cultural heritage, natural and ecological zones with safeguarding regimes.

The 'written policy statement' contains the conditions of any activity in the area from the point of view of the conservatory authorities, space and functional regulations, and the categorization of all areas of utilized territory.

The suggested layout and some other more detailed design guides were used during the preparation of a special drawing entitled 'Protection of image of town and landscape in statutory town plan of Telč' (Fig. 3.3). It represents a minimal success in the integration of cultural heritage protection and urban planning achievable within the framework of the prevailing Czech legislative environment.

It is important that the General Plan contains *strategic conditions* to preserve the historical monuments and natural values in a town. It is also necessary to establish a *specific conservation policy* for each part of the territory for the plan area. This policy should be independent of fashionable waves or changes of approach (even scientifically justified), and should assure a continuity of interventions, especially in large and important architectural ensembles (see the section below, 'Management and regeneration action'). Specific conservation policy in a limited territory should be fixed after the evaluation of its cultural heritage potential and in close interrelation with the *urban planning policy of the municipality.* Evaluation of the *composition characteristics of a town* should form an integral part of these strategic conditions. There should also be areas set up to cover different urban composition variants related to historical and natural values. These will have to be considered in the preparation of 'urban plans'.

The methodology to be used for the preparation of the General Plan is the subject of ongoing discussions. More significantly, the present opinions of the conservation authorities and organizations about the possibility of elaborating a plan in conceptual way are quite sceptical. They point out that such a document would not be accepted by designing architects/urban planners if they were not to be responsible for its compilation. Nevertheless, if the plan is to be regarded as an urban planning document, such attitudes will need to change. This will allow for the complex issues to be observed through methodological guidelines.

In this context the methodology for the General Plan should define:

- the elements of heritage values of a territory (of spatial, area, line or solitary character);
- relations between these elements (axis of composition, axial views, historical and immaterial relations);
- evaluation criteria of the elements of heritage value in a territory and of relations between them;
- categorization of the elements of heritage value of a territory (primary importance, secondary importance);
- evaluation criteria of the impact of the proposed projects upon heritage value of an area;
- opportunities and conditions for the preparation of a specific conservatory policy for a municipality.

One important consideration for the plan is the idea of dividing the areas concerned into categories. Three categories have been proposed.

- *Area which determines the character of a historical town* – a territory submitted to the strictest cultural heritage protection where interventions disturbing its historical character (image), historical substance, and cultural and historic identity have to be eliminated (including providing harmony with natural elements and its relation to the geomorphology of the territory).
- *Area which accompanies the character of a historical town* – a territory submitted to strict cultural heritage protection targeted to the protection of its basic image, traces of urban development and historical activity and its continuity with other parts of the town.
- *Other area* – a territory with only basic cultural heritage protection where a complex transformation or development is possible.

Special regulations (spatial and functional) for these areas are to be declared. These areas could be divided into smaller urban elements with more detailed regulation. There are further classified individual elements (objects, areas, etc.) in all territories.

The same methodology, i.e. the clearly defined zoning and relevant measures, should be adopted in land-use and protection planning tools for a cultural landscape or countryside. In the Telč micro-region an attempt by the Ministry of Culture to designate a new cultural heritage landscape zone failed due to strong opposition expressed by the local authority. However, the proposal had not been clearly formulated with regard to zones of differential heritage importance and supporting measures.

The General Plan does not usually have to contain specifications for building interventions for every object. This specification has to be determined before issuing a construction permit according to the conditions formulated by the relevant regulation. This approach enables the preparation of the plan within the formulation phase of the municipal land-use plan, as has been done for Telč. Policies on building interventions concerning heritage objects can be specified either by a more detailed safeguarding plan for a zone, or by an *ad hoc* method after the elaboration of detailed archaeological investigation and evaluation. It is thought to be too dangerous to specify the building intervention possibilities without the benefit of a building archaeological survey. To elaborate such a survey for a historical part of a town (even for a small one), is very expensive and takes a long time, it could thus delay the preparation of the Safeguarding and Enhancement Plan.

According to the current Czech legislature it may be possible to incorporate the concept of integrated plans into the planning and statutory documents of a historical town (by reference to the Urban Planning and Building Code No. 50/1976). It is thought that the General Plan could be developed in the context of the land-use plan, in much the same way as the territorial system of ecological stability used to be prepared (i.e. having a special status). However, this will require a modification of the law on the care of monuments. If this kind of change can be achieved, the juridical consequences would be really strong. It would assure a more forceful protection of cultural heritage values in the territory even before the preparation of the general urban planning documentation on the corresponding level and before issuing the municipal by-law. On the other hand, the reduction of this document to the level of a mere land-use plan would greatly reduce the urban planning values of the safeguarding plan and would reduce the preventative conservation mechanisms in the surrounding parts of the town. These areas are very vulnerable and at the same time they are exposed to the strongest transformation efforts.

Plan for the safeguarding of cultural heritage

The basic objectives of this second step of developing a Safeguarding and Enhancement Plan will have a more detailed emphasis. This type of plan will include:

- the safeguarding and conservation of the historical cultural values of each object before or during the process of preparation of urban planning documentation;
- the preservation of the context and the continuity between natural, landscape and urban elements in the territory;
- extensive rules to allow development of the historical part of a town, as an exceptional ensemble in its unity and coherence, and to intervene in individual cases,
- the constitution of an urban planning document which would become a compulsory part of the regulation plan of the town;
- the identification and recording of objects of cultural heritage value, above all those which are not objects of conservation protection;
- the elaboration of a record of the physical condition of each building ('passport') and the preparation of regeneration plans.

The content of the Safeguarding and Enhancement Plan is described in detail elsewhere (Drdácký, 1999). It generally includes:

- the main objective for the plan, especially the limits for *'determining', 'accompanying' or 'other' areas* (see above) and their sectors (smaller areas within these areas based on the categorization identified above);
- an analysis of the area including the definition of urban and architectural elements of differential quality;
- analysis and conditions for urban development, namely for housing and other functions in association with municipal urban policy with a view to:
  - encouraging the social rehabilitation of historical objects according to the requirements of local inhabitants and taking into account their economical wealth,
  - enhancement of the architectural heritage to make the town or its centre more attractive,
  - halting the depopulating of the city centre,
  - developing economic tools against speculative risks (such as land price inflation),
  - maintaining the possibility of developing the town and enhancing the monuments).

The main part of the Safeguarding and Enhancement Plan should accommodate the aims and the rules for conservation and enhancement, for example:

- exposure of historical monuments and increasing the legibility of historical traces in a town;
- the conservation of sites – including panoramic views and regimes exceeding the town area;
- transformation of historical monuments – based on the removal of parasitic elements associated with heritage objects;
- the return of housing to the city centre – the main priority of the plan;
- architectural provisions with criteria and classifications of objects and areas;
- the definition of rules for the regime of the Safeguarding and Enhancement Plan (for example: conservation, reconstruction of original state, restoration, new use of objects or elements, modification, conservation of elements, indication of traces);
- regulations and rules for architectural design (for new building and alteration of existing objects);
- rules for new construction – including the criteria for the integration of new objects or elements;
- terminology – including definitions, regulations for safeguarding and enhancement of non-built elements, skyline/horizon elements, scenic views and urban planning provisions.

The aim is not to change the economy of the historical area but to *reinforce its role in the city and to reinforce selected functions, which are arising from the town's historical substance*. The interventions can include: the *freeing up of inner parts of blocks* (partial demolitions); the *completion of blocks* (to delimit an area to be built-up); the *opening of public spaces or the creation of new ones* (to display city walls, new access to buildings, roads, modifications of city interiors), etc.

The methodology for the Safeguarding and Enhancement Plan should include:

- elements of heritage value associated with objects in the territory;
- evaluation criteria for such objects;
- the categorization of elements of heritage values in a territory (objects and areas);
- evaluation criteria for interventions;
- the potential to formulate specific conservatory policies in relation to the use of single objects.

As mentioned in the case of the General Plan, the preparation of the Safeguarding and Enhancement Plan should take advantage of the building archaeology surveys or an evaluation of rural architecture.

The plan (usually to the scale 1/1000, 1/2000, or 1/5000) can be successfully consulted and approved as a statutory zone land-use plan or a regulation plan. If an approved General Plan exists, the safeguarding plan could be sufficiently prepared as urban planning datum only, because the public interest can be assured by this General Plan and municipal interests by an urban plan at the level of regulation plans. This process should be assured by modification of the Law on the Care of Monuments and also by the Building Code and the issuing of a municipal by-law.

This combined variant does not contradict European standards. It can satisfy every participant, including the public, and can be sufficiently quick and operative for the preservation of important cultural heritage values in an area. It would give enough time to allow the elaboration of the scientific data and prepare the safeguarding plans at the level of regulation plans and it would be more independent in individual aims and on political and other tendencies at the moment.

Design briefs and guides

During the work on both general/safeguarding plan guidelines and the land-use plan for Telč, attempts were made to use design briefs or guides for new development and for the restoration of the built heritage. They were subject to strong criticism by the co-operating architects as well as the heritage authorities. It seems that this tool in the form of obligatory guides represents a very sensitive matter and needs to be resolved on several levels.

Nevertheless, a catalogue summarizing architectural elements of the region, the city and surrounding villages, together with recommended formal as well as material examples for reconstruction, repair, conservation, remodelling or even new construction is very useful and desirable. Only such an approach could help to maintain the regional diversity of the world heritage built environment of Telč.

Management and regeneration action

The role of national and local authorities in heritage management

It has been recognized that historic cities need a specific conservation policy for their safeguarding and enhancement. Such a policy should be independent of fashionable waves or changes of conservation approaches and should assure a continuity of interventions in valuable architectural ensembles. This can be illustrated by examples from Telč where an analytical approach was adopted in the 1950s during massive restoration works (Fig. 3.4). Fig. 3.5 shows the result of recent restoration action which does not comply with the original policy and tries to reflect modern approaches to the conservation of all architectural development layers

Fig. 3.4 House No. 15 before restoration (1953) – left, and after restoration in late 1950s – right. Example of the analytical approach adopted for restoration of Telč houses in the 1950s.

Fig. 3.5 House No. 31 in about 1900 – left, and after restoration in 1999 – right. Example of a contemporary approach which has disturbed the architectural beauty and clarity.

present on the building. Unfortunately, House No. 31 lost the essence of history, architectural beauty and authenticity. It is a typical example of the sad consequences of an inconsistent conservation policy, as well as of unsuitable heritage management practice.

There is a lack of co-ordinated activity from heritage protection officers, state authorities and town governing bodies in Czech heritage management practice. Heritage protection officers represent two independent bodies: (State) Heritage Institutes, which are only entitled to elaborate professional opinions and recommendations, and Cultural Departments of District Offices, which issue the approvals necessary for building permission. Building permissions are issued separately by (State Government) Building Offices, independent of the local governmental (elected) bodies. Municipalities can establish town architect offices, which have very limited legal powers in relation to the building permission process. Under such circumstances, all interests may be created only by means of legal standards and documents. This explains the need for new integrated conservation-planning planning tools (see above).

The Czech Government has approved a so-called *Programme of Regeneration of Historic Towns*, which is a useful tool for the promotion of revitalization efforts in historic cities. In this programme, the role of local governments is strengthened, because the individual programmes are approved by elected bodies and the documents create a basis for negotiations of local committees allocating financial assistance to the owners of heritage objects.

Funding of conservation and regeneration efforts in historic centres

The new Czech law concerning 'foundations' requires a minimum basic fund of 500,000 CZK. In practice a foundation can only use money above this figure, but it is difficult to attract sufficient further funding. This prevents the creation or existence of any local foundations in historic cities or regions. Of course, there are several state programmes for financial support for the conservation or the safeguarding of a built cultural heritage. Their effect is both positive and negative. The main disadvantages are connected, on one hand, to the administrative measures and awkward procedures, which can cause a very late release of allocated money. In many cases this results in a waste of resources due to a necessity to spend the money by the end of the current year, or low quality repair works due to climatic conditions unfavourable for outdoor construction. On the other hand, the prescribed bidding and contracts with official construction companies raise the costs to a level that is in many cases unaffordable for a number of owners.

For these reasons, some private owners in Telč prefer to repair or maintain their houses without any financial support from the state and are even proud to do so. Fig. 3.6 shows one such house. Such an approach helps to retain the authenticity of a cultural object. On the other hand, the owners need special and continuous advice concerning suitable technologies, procedures and materials.

Fig. 3.6 House No. 50 in about 1908 – left, and in 1999 – right. Example of 'soft' and inexpensive maintenance.

Environmental management

Traffic is one of the most difficult problems in the majority of small historic towns in the Czech Republic. Car ownership has increased significantly in recent years. In Telč the problem has not been sufficiently resolved. A preferred policy for car parks is to have them widely dispersed within the area of the city, while restraining their size. New car parks are being designed in the form of a park of mature trees in order to reduce the visual impact on the historic environment. It would be desirable to revitalize the existing railway system to ease the problem of car usage, but, at present, action is still awaited.

Transformation of heating systems and improved fuel supply of gas and electricity has helped to decrease air pollution in the city. Research facilities in Telč are involved in an international programme concerning the deterioration of heritage due to pollution (REACH) and as a result the air quality is continually measured. In 1999 a new sewage network and a purification plant were built. Unfortunately, all these environmentally necessary works have damaged much of the archaeological heritage in the city. This fact further emphasizes the need for integrated conservation planning mechanisms.

Tourism and heritage management

Most countries in Central and Eastern Europe do not have the ideal conditions for recreational tourism. On the other hand, a great amount of well-preserved cultural monuments and the density of historical cities attract many cultural tourists. In past years cultural tourism was often strongly combined with shopping, which is one of the hidden aspects of tourism. Even today, a visit to the Czech Republic presents a convenient possibility for the cheaper purchase of quality crystal or other goods, and thus is a good destination for many tourists.

The opening of the frontiers did not, however, bring the expected increase in tourism to the small cities. On the contrary, in Telč, for example, a decrease in visitors was recorded compared to the years before 1990. This was caused by an increase in living expenses, an expansion of recreational alternatives abroad (skiing in the Alps and recreation at sea resorts for instance) and also by the loss of leisure time among many of the new entrepreneurs. The spectrum of foreign visitors has changed as well because other countries of Central and Eastern Europe are experiencing a similar situation and Western tourists are still only starting to discover these countries. Telč, an example of a small World Heritage City, welcomes approximately 80,000 registered tourists each year, while estimates of the actual number of visitors suggest that the figure is nearer 150,000 visitors (Fig. 3.7). This volume is about ten times smaller than is common in similar cities of Western Europe and so far has not caused too many significant problems for the historic centre, although the need to manage tourist activity is recognized. The economic contribution of tourism to the city of Telč is very low, partially due to the small number of visitors and their poor financial status, but also because of an inappropriate tax system.

Tourism development

There is practically no experience available to deal with the speed of tourist development in the countries of Central and Eastern Europe. The fact is that a wave of unplanned and expanding tourism could have potentially destructive consequences. With this possibility in mind it is thought to be necessary to investigate the potential to develop tourism in the interior of the historic city and to prepare a strategy for this purpose through planning and management tools. A map of this potential ought to be a part of the development plan.

Tourism demands a special infrastructure, which in other areas of a city's life may not be relevant. Such an infrastructure can stay idle for the majority of the year due to the seasonal nature of tourism and requires costly maintenance. To overcome this problem the concept of the so-called 'distributed hotel' has been developed in Telč. This provides accommod-

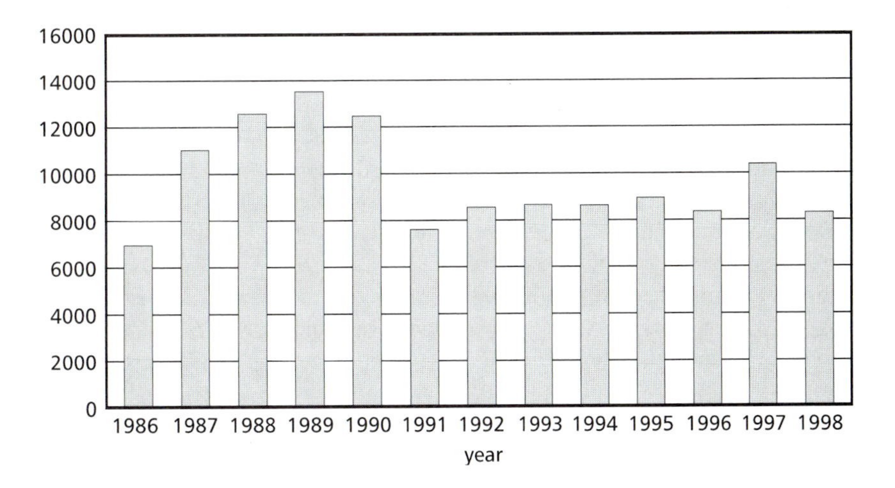

Fig. 3.7 Review of 'visitor' development in Telč.

ation in historic houses on the central square through the effective use of unoccupied spaces, serviced by the resident owner. This provides a beneficial income for the city. There are also a few small hotels (around 25 bed places) which incorporate dining facilities. This helps to preserve life in the central core by offering tourists an attractive service that does not disturb the existing scale and rhythm of the city's life.

Perhaps the worst consequence of tourism for the infrastructure is the remarkable change in the composition of the businesses in the historic core. Tourist shops have gradually driven out facilities that served the local population. This has had a detrimental impact on both local residents and the businesses that served them. This is especially true for the 'arched square' which used to house a diverse range of shops creating a large open department store, easily accessible during bad weather conditions.

Tourism has thus had a negative impact, particularly on the natural life and tradition in the historic core of the city. Entrepreneurial activity, usually led by individuals who do not live in the city, has been exercised for personal profit with no regard to the protection of public interests. There is a danger that this may transform the city into a cheap attraction (usually represented by markets, tourist shops and festivals and by the theatrical illumination of monuments). The historic part of town therefore had an artificial renaissance. This situation brings a number of other problems, particularly out of season, when the historic parts of town are empty for long periods of time.

In order to overcome those problems, a complete knowledge of the city's interior development potential and a well-prepared strategy is vital. In regard to this notion it is necessary to support such activities that will satisfy the needs of tourists while demanding the participation of inhabitants, (by, for example, the 'distributed hotel'). There is, therefore, a need to conserve not only monuments themselves but also the whole atmosphere of the city.

Tourist management policy

A general policy is to be developed in the region to prevent the creation of mass tourism centres and villages or enormous hotels. The activities are aimed at keeping the scale of tourism activity within the thresholds of sustainable capacity in historic centres in general. There is an effort to disperse cultural tourists onto a larger area and offer them places to visit other than monuments inscribed onto the World Heritage List of UNESCO. To assist this process, a travel bureau, 'Czech Inspiration', was established, which has united tourism management efforts for three World Heritage Cities together with three other historic centres.

Another necessity for successful tourist management is educated experts. In 1997 a specialized college aimed at managing travel and tourism issues was set up in Telč. Many potentially consequences of tourism on the historic city are still to be explored. In this context special inquiries and focused research are being undertaken to be able to define the problems and find appropriate solutions from a wide range of social, economic, technical, cultural and political viewpoints.

Cultural role of the historic centre

One of the basic symbols of historic towns seems to be that of the home, a symbol of a safe place for natural life. It contains not only spatial aspects of the context between the site and its environment but also aspects of its natural life, the rhythm of life being the most valuable.

Telč tries to conserve this intimacy of the historic town life in harmony with its intimate spaces. However, in reality there has been a failure to prevent the city from becoming something of a theatre stage set. But the symbol of the home and a natural place for life is still strongly felt, not only by the town citizens but also by the town visitors. This assists in protecting the heritage by recognizing a 'common wealth'. The visitors or guests understand this symbol in a general sense. For the citizens it has a very actual and real meaning and plays an important role in binding home life with communication networks between the citizens and the town governors, citizens and tourists, with relations of differential polarity and content.

Conclusions

The practical problems associated with living in a World Heritage City can only be solved through continual communication between the heritage protectors, city planners and managers and the citizens to and the negotiation of a *common understanding of problems*.

Most of the resident and business population of Telč can identify with the heritage that they utilize and they have a good will to preserve and maintain it. But there is also a need to live and work within the heritage according to contemporary conditions. Our heritage has not been created for the purpose of admiration, it was built to be used. In many cases there are no reasonable objections to prevent the utilization of heritage for a valuable modern life. There is a need for a *clearer defined philosophy of heritage protection balanced against utilization requirements*. For this purpose the official protectors must study continuously, and in detail, the principles of life in heritage sites. The plan proposals for integrated conservation mechanisms will assist this process if there is the will to adopt widely-accepted European standards for the management of the immovable heritage.

At the present time, there is a lack of *educational activities for a broad audience* that offer, on a sufficiently high level, a useful knowledge concerning the possibilities for using the heritage in a sustainable manner. More specialist knowledge of maintenance techniques, acceptable modernization changes, building refurbishment methods and financing possibilities is needed. The protectors and users must know and speak the same language to be able to understand each other. This, in turn, will require the continued support of broad international collaboration, exchange of experience, and even research.

In this context, it is necessary to create a database *of technological procedures of heritage skills and crafts* and to support the publishing of the saved knowledge and its distribution into all historic sites in the Czech Republic. Similarly valuable will be a database of *failures, errors and bad examples* connected with heritage activities. For Telč the process of developing a sustainable framework for this important historic centre is just starting. The important consideration is that the process has begun.

References

Council of Europe (1995): *Recommendation No. R (95) 3 of the Committee of Ministers to member states on co-ordinating the documentation methods and systems related to historic buildings and monuments of the architectural heritage.*

Drdácký, M.F., Macháček, J., Rákosníková, V., Bláha, J. (1998): *Documentation system, methodology of economic and non-economic assessment and lines for safeguarding plans of historic towns with a demonstration plan of Telč (in Czech)*, ITAM, Academy of Sciences of the Czech Republic, Research Reports.

Drdácký, M.F. (1999): *Methodology for cultural heritage safeguarding and enhancement plans in small historic towns and micro-regions*, Proceedings of the 5th International Symposium OWHC 'Tourism and World Heritage Cities – Challenges and Opportunities', Santiago de Compostela, 15 October 1999.

Matoušková, K. (1998): Comparative analysis of territorial protection of cultural heritage – new tools, legislation, foreign experience (in Czech), Official Paper (unpublished).

Pickard, R.D. (ed.) (2001): *Policy and Law in Heritage Conservation,* Conservation of the European Built Heritage Series, Spon Press, London (see Chapter 3: Czech Republic by J. Stulč).

Syrová, Z. and Syrový, J. (1997): *Building archaeological and urban history survey of the National Park 'Podyjí' – an example of GIS System of Urban Stability* (in Czech), CD ROM, Ministry of Culture of the Czech Republic grant project 853/2.

Thornes, R. and Bold, J. (eds) (1998): *Documenting the Cultural Heritage*, Getty Information Institute, Los Angeles.

Erling Sonne

Ribe, Denmark

Introduction

With its first settlements dating back to the seventh century, Ribe is the oldest city in Denmark, and also one of the best preserved. It is situated in the south-west of Jutland in the immediate vicinity of the tidal flats known as the Jutland Wadden Sea, which – in conjunction with the German and Dutch tidal flats – have been designated as European Union bird-protection areas.

The city is situated on elevated land, where marsh and moor meet the so-called 'Geest'. Throughout history, the city has been situated at a height, which has provided a fair amount of protection against the recurring floods in the area.

Ribe was originally formed as a Viking settlement founded north of the River Ribe at a point where the river was very narrow. After the introduction of Christianity in Denmark, the Christian quarter was founded by the southern part of the river, and a cathedral was erected here during the thirteenth century (Fig. 4.1).

At an early stage, the city constituted a central place of commerce, with the east–west waterway connecting with the north–south main road. At the same time, the erection of the cathedral provided Ribe with the status of a cathedral city, and during medieval times the city boasted sixteen convents and churches. All this gave the city its important position; a position which lasted as long as Ribe was the main gateway for imports into and exports from the kingdom, and until the Lutheran Reformation weakened its importance as a religious centre.

Throughout history, the landscape and settlements of the Ribe area have been aligned on a north–south axis. Indeed, when the highway and the railway were founded in the late nineteenth century, the infrastructure connected landscape and settlements along this axis and created a direct and very important commercial link to countries south of the border.

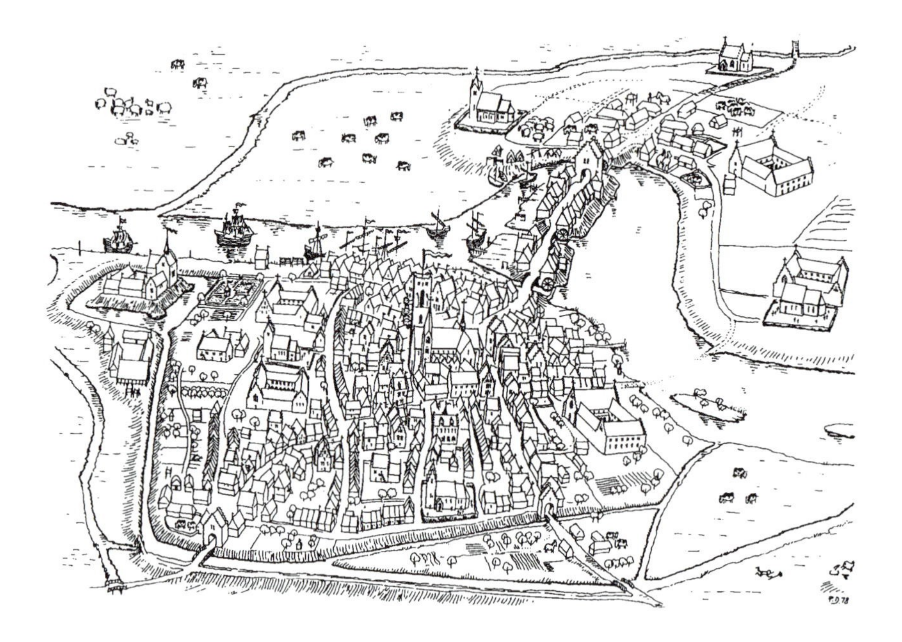

Fig. 4.1 Historical map of Ribe – 'The Double Town' with the Viking settlement and the Christian town on separate sides of the river.

Ribe city and its surrounding landscape display a very harmonious interplay. This is mainly a feature of the fact that the city is situated on small sand banks in a flat, sparsely forested river valley with extensive meadows, bogs and marshes. As a consequence, the city's potential for growth has been limited to the relatively small areas on the sand banks. For this reason, Ribe became a densely built-up city with very clear and precise demarcation between itself and the surrounding countryside and landscape, among other things, signified by city gates.

From the mid-seventeenth century, the River Ribe gradually lost its importance as a trade route, and Ribe went into stagnation. This fact was aggravated by the delineation of the frontier after the war in 1864. The new border was established along the King River, but did, however, make a slight turn south of Ribe. The city remained Danish, but lost its entire hinterland towards the south and east. When this hinterland was returned to Ribe in the 1920 Reunion following World War I, the city of Esbjerg had long since taken over the role as Denmark's main gateway to the west.

The policy and planning framework

The physical condition of the historical environment

Ribe's special position in the landscape and the topographical conditions mean that the city has limited potential for growth. This, in conjunction with historical parameters, has occasioned the city's development to occur in the form of satellite towns that are clearly separated from the historical core of the city by green meadows and wetlands.

This natural foundation still constitutes the basic element of planning. None the less, it is evident that since the 1960s, there has been an increasing demand for new areas to be used for business purposes in connection with the historical city centre. This has occasioned a weakening of the city front in several places due to the erection of poorly adapted buildings. This means that demarcation of the city, towards the south especially, but also at the northern approach road, is now discordant and without structure; there is no reflection of a clear position in terms of planning and architecture (Fig. 4.2).

During recent years, the demand for developing new growth areas has increased continually, but at the same time, an increased awareness has emerged among local authorities and the public regarding the necessity of creating a balance between preservation and the extension of the historical environment. In more specific terms, this means that a future housing development area, to be used for new housing development throughout the coming 15–20 years, will be situated with a view of the old city centre, yet separated by green belts. This will maintain the demarcation of the city while placing the area in close contact with the surrounding countryside and nature.

The development of businesses, trade and retailing in Ribe presents a problem, as landscape features mean that, within the city centre, only very few places exist where it is possible to procure new sites, particularly for retailing purposes. One solution in terms of planning would be to identify a new development area in the northern part of the city centre; however, this would also entail a further weakening of the northern city front and obscure the clear demarcation towards the surrounding countryside. Consequently, any development works will place great demands on the design of the area in terms of planning and architecture.

However, the historical city centre still plays a role in the development and expansion for housing as well as business purposes. Thus, opportunities,

Fig. 4.2 Bridge and meadow seen from the Ring Road – nature is a master planner and helped create the clear demarcation between countryside and city.

albeit limited, remain for recreating and finishing old historical buildings and for demolishing existing, worn-out and architecturally weak buildings with a view to creating space for new buildings. In turn, this will assist in developing and enhancing the historical environment.

The integration of urban planning and the historical built environment

The idea of preserving the medieval city centre originated as far back as 1899, when the first association for the preservation of historic buildings, called the Ribe Tourist Association, was founded. By example, in the 1950s, the Ribe Tourist Association began to buy properties 'worthy of preservation' with a view to restoring them and then selling them. One of the main objectives of this association was, and remains, to continue work to secure and preserve historic buildings within the city centre.

In 1963, the Ribe City Council adopted a so-called 'Preservation Declaration' for all houses within the city centre – a total of approximately 500 – which meant that all external changes and alterations to these buildings had to be approved by Ribe's local authorities (Ribe City Council, 1963). This was to ensure the continued use of materials and craftsmanship in accordance with historical traditions. By means of a 1989 local plan, this preservation area has been extended by two adjacent housing developments from the twentieth century, comprising a total of approximately 200 houses.

The general planning tool currently available is the *Local Authority Plan for Ribe 1998–2009,* which describes the overall objectives and frameworks in terms of policies and planning for the municipality as a whole (Ribe Municipality, 1998). On one level below the local authority plan there are the local plans. These contain more specific guidelines and provisions on smaller geographical areas and specific projects.

As a more active development in relation to the guidelines and policies of the local authority plan, a discussion document called the *Ribe Urban Plan* was prepared in 1999. Among other things, this plan addresses issues regarding preservation and planning and connects them to specific development plans and action plans. The objective of the Urban Plan is to address a series of current issues and problems with respect to the development of Ribe from a holistic point of view. On this basis, principles are established on Ribe's development during the decades to come. This plan has been prepared by the Ribe local authority with the support of a Panel comprising twenty people representing a series of organizations and associations with an immediate interest in the development of the city. The plan was subsequently subject to a period of public hearing, following which it was finally adopted by the City Council.

This plan has the following main themes: retail structure, traffic structure, preservation and urban renewal, streets and squares, parks and avenues, and city demarcation. It describes the current issues and problems to be addressed and decided upon. In addition to this, the plan contains political objectives for individual main themes and describes the areas to be addressed during the years to come. Finally, these areas are followed up by descriptions of a series of projects that are to form part of the future work of the local authorities.

The objective of the Urban Plan, which is aimed at citizens, politicians and planners, is to direct attention to the series of issues and problems which will require resources in terms of planning and finance in the near future. The objectives of the Urban Plan are:

- to secure the historical and cultural environment of Ribe city;
- to enhance development of Ribe city as an active and contemporary centre of trade, business, tourism and leisure;
- to reduce traffic in the city centre with a view to protecting the urban environment and old houses;
- to improve conditions for city-centre residents, visitors, retailers, and businesses by ensuring an effective traffic structure with good public transport;
- to ensure appropriate and effective use of parking opportunities in and around the city centre with a view to keeping parking within the city centre to a minimum;
- to preserve and improve the well-defined city demarcation against the surrounding wetlands;
- to ensure that development and preservation of Ribe is carried out with a view to creating quality and a good urban environment by continuing the historical traditions of Ribe city;
- to ensure appropriate care for and renewal of the city parks and avenues.

Even though the plan contains several main areas, the greatest issues of conflict concern the narrow and compact medieval urban structure, and its inability to absorb the ever-increasing traffic, especially during summer. Thus, there is a clear conflict of interests between the different roles of Ribe as a residential area, commercial centre, tourist attraction, and historical centre; a conflict which can only be resolved by weighing individual interests in relation to the established objectives.

As an example of an overall planning tool, the National Forest and Nature Agency (NFNA) (the government agency responsible for building preservation matters), under the auspices of the Ministry of Environment and Energy (MEE), has had the opportunity of preparing so-called *Preservation Atlases* with Danish local authorities. This is undertaken through a methodology known as *InterSAVE.*

The initial drive for developing this system was the signing of the Granada Convention in 1985 (and ratification in 1987). The convention stressed the importance of 'integrated conservation' and the first steps towards developing this approach were taken in 1987. Since 1990 the InterSAVE system (Survey of Architectural Values in the Environment) has been operated through three phases. Phase I involves a 'Preliminary Investigation' of an area which is followed by Phase II (the 'Fieldwork'), to provide additional information on 'developed structures' (dominant architectural features, building patterns and selected urban settlements) and more specific information on individual buildings. The information from the two phases is brought together in Phase III through the publication of the 'Preservation Atlas' (MEE/NFNA, 1995).

A Preservation Atlas was developed for Ribe in 1990 (Ministry of Environment, 1990). It describes the special historical and topographical conditions that have caused the building patterns of Ribe to become so distinctive and special. The Preservation Atlas also indicates the characteristics of individual periods, and thus indirectly indicates the urban elements and building elements that can be developed further in terms of quality. Based on the different phases, the Preservation Atlas for Ribe describes developed structures on the basis of topographical surveys, historical analyses and architectural observations. In addition to this, architectural values are described on the basis of six different conditions. In turn, this forms the basis for a summary of the general preservation value of individual buildings. Thus, the Ribe atlas provides a 'modern' way of combining the traditional view of preservation, which concentrates on individual buildings, with a progressive planning view, and therefore provides a more holistic approach to the management of the urban environment.

In order to be able to use the atlas as a tool in connection with preservation work, significant elements of this atlas have been incorporated in the *Local Authority Plan for Ribe 1998–2009* (as discussed above) and have been used as the basis for preparing guidelines and more specific action plans.

The use of design/development briefs and the integration of new development within historical settings

As a natural consequence of the strategy for preservation work in central Ribe, a series of principles regarding urban design have been developed with a view to ensuring quality with respect to both preservation and development.

Homeowners' requirements today regarding dwelling areas, lighting, sanitary installations, maintenance, etc., may be incompatible with the layout of an old town house, especially if permission is not granted for changing or adding to the existing house. With a view to resolving the dilemma between preservation interests and the desire for additions, the local authorities have prepared guidelines on how to manage additions and extensions in order not to destroy architectural values.

In order to maintain the overall impression of the streets and architectural structure, the main principle dictates that new buildings must be placed in alignment with other buildings in the street and that they must be adapted to the surrounding buildings in term of both proportions, horizontal and vertical divisions, etc. Furthermore, in principle, extensions and annexes must be designed as lateral additions or back buildings to existing buildings. By doing so, acceptable solutions that meet current requirements and demands are created.

In addition, the Local Authority Plan for Ribe 1998–2009 has been among the documents serving as a basis for a series of guidelines providing more detailed explanations on the specific counselling and administration approach of the local authorities. The objective of these guidelines is to ensure that details are protected on the basis of the principle that if details disappear, the totality is watered down.

Application of these management tools is mainly connected to the renovation and maintenance of individual houses; sign displays, lighting, and façade design with respect to shops; erecting new buildings in historical environments; city parks; avenues; block areas; and, last but by no means least, public sector design.

Renovation and maintenance

With respect to the renovation and maintenance of individual buildings, a set of guidelines entitled 'Good Advice on Old Houses' has been prepared, providing the public with guidelines on how to maintain and renovate the old houses in central Ribe (Ribe Municipality, 1997). These guidelines provide detailed descriptions of how all individual building parts, such as gables, windows, chimneys, doors, etc., should be designed in relation to the architecture of individual houses. For example, with respect to windows, the guidelines include specific descriptions of the design of windows; the dimensions of individual glass panes in relation to the overall outline of the window and bar dimensions. As an extension of these guidelines, detailed working drawings have been prepared for a large number of building elements (Fig. 4.3).

Work is currently being carried out with respect to the issue of paint types and colours. The old traditions are based on whitewashes and natural colour washes; however, during recent years, silicate paints have become an alternative, the use of which is being debated with respect to the treatment of old houses. For example, work is being carried out on establishing a colour template based on old colour washes which can then be transposed in terms of colour codes of silicate paints, and on preparing guidelines within this area.

Signs and lighting

The use of design briefs for signs, lighting and façade designs with respect to shops within former residential buildings is based on the principle that the building should appear as a whole. Thus, by this approach, the original

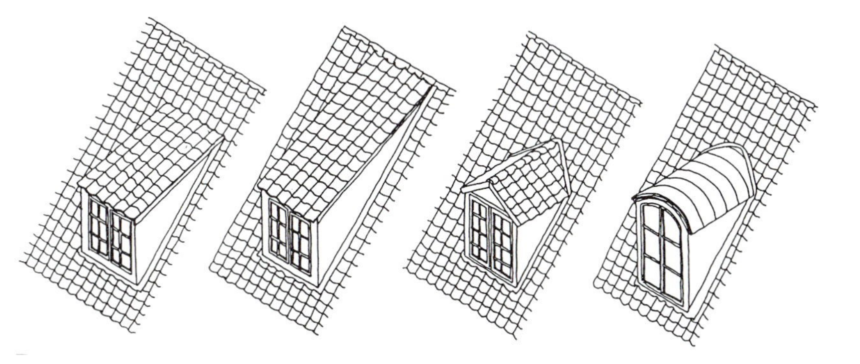

Fig. 4.3 Principles for gables – it is important to apply correct principles which match the house.

historical features of houses can be preserved, creating coherence between individual floors both vertically and horizontally and a balanced interplay between the 'house', signs and lighting.

During the 1960s, many of the original façades were significantly altered and to a large extent ruined. Now, most façades have once again been returned to a more authentic appearance in accordance with the principles behind the preservation efforts.

Today, the main points of the design principle are:

- that signs should now consist of individually designed wrought-iron hanging signs and of carved lettering on wall surfaces;
- that the concept of night-time marking is used instead of sign spotlights as a basis for creating indirect lighting behind individual carved letters in order to achieve a muted corona effect;
- that the dimensions and design of shop windows must be adapted to the character of the house and the windows of other floors;
- that no solid canopies be used, only cloth awnings.

Erecting new buildings

During recent years, a great deal of consideration has been given to the principles to be applied to the development of new buildings. Generally speaking, no authorizations are given for demolishing buildings, unless it can be proven that the building in question is of little significance in terms of architectural or historical importance, or that it is in such poor condition that any attempt at restoration must be considered unfeasible. Nevertheless, when opportunities are provided for erecting new buildings – either as individual houses or as interconnected larger structures – the local authority plan incorporates guidelines for the overall building structures and more detailed provisions providing frameworks for their volume and adaptation into the neighbourhood in terms of scale.

Balancing the relative importance of the more detailed and specific principles for building design and materials is a very complex issue and cannot be described in unambiguous terms. The extreme ends of the scale range from constructing buildings which stand out prominently from the existing buildings in terms of design and materials, thus signalling a different era, to building exact replicas of old houses, making it impossible to distinguish between old and new.

In Ribe the traditional architectural expressions, in terms of design and materials, are being reflected in new design approaches and modes with a view to having houses appear as individual, independent town houses. An example of this would be the new 1995 holiday accommodation – *Ribe Byferie* (Ribe Holiday Town) – comprising 96 flats constructed in such a manner as to constitute a new, clear demarcation of the city centre towards the south (Fig. 4.4). The actual construction plan of Ribe Holiday Town is structurally identical to the plan and layout of the buildings at the old city harbour, and the recurring architectural theme – the gable houses – hails from Ribe's distinctive old seventeenth-century gable houses. None the less,

Fig. 4.4 Ribe Holiday Town – this development heralded the creation of a new, clear demarcation between the city and the surrounding countryside.

the proportions of the houses, their positioning and their distinctive use of traditional tiles and bricks, clearly signal that these buildings were erected in the 1990s.

The use of design briefs faces a series of general problems as technological developments mean that a number of new products and techniques available 'look like' old types of materials, but are easier to work with, cheaper and often easier to maintain. The most typical examples include plastic windows with bar dimensions identical to those of wooden frames, the use of concrete tiles and bricks rather than clay-based materials, external insulation of façades, synthetic joint fillers rather mortar fillers, new types of paint to replace whitewashes and colour washes, etc.

In this respect, it is important to maintain that materials should not only *look* right: the craftsmanship and materials must remain in keeping with individual houses. So far, it has been possible to uphold this principle in Ribe, and only in a few areas – for instance with regard to the use of silicate paints – has a shift in the use of materials occurred.

The point of departure selected in Ribe is in some ways problematic, since modern requirements regarding accommodation and buildings, in conjunction with building profitability, mean that these projects involve strong desires on the part of developers to maximize construction volume. In other words, a certain level of the floor area is required to make a scheme profitable. This means that compromises are sometimes reached which lower the quality of the final appearance of buildings.

Green areas

The historical green areas of Ribe are very influential on the city's environment. The medieval city centre of Ribe is characterized by small, narrow and densely built-up streets with narrow courtyards, alternating with large green areas within housing blocks. The parks with their towering old trees and the old avenues constitute a prominent and significant element in the structure of the city and are legacies of a period when green areas were largely experienced in passing or on a stroll.

Today, we prefer to view the parks and avenues as places to pause – somewhere for quiet reflection. Consequently, future use of the parks must constitute a balance between preservation of the old atmospheric historical features and a more forward-looking use providing inspiration for pause and experiences in an inter-play between parks and water. With respect to the avenues, the original tradition of using avenues in connection with the establishment of and changes to approach roads must be continued, and trees should be replaced whenever the need arises.

In several places in central Ribe, green areas within the confines of ensembles have been reduced in size due to the partial use of these areas for buildings and parking spaces. In order to be able to curb this development in future, principles governing the use of green areas have been incorporated in the Local Authority Plan for Ribe and in the Ribe Urban Plan. However, an element of pressure remains with respect to establishing parking spaces in parts of the blocks in order to alleviate parking problems in the narrow streets.

Public spaces

The use of design and development briefs in connection with public spaces is an important and vital factor in the preservation of the historical environment. The medieval feel to Ribe is very much reflected by curving streets that open up into market places and squares. Throughout history, these market places and squares have, in Ribe, been assigned special functions and usage, such as fish markets and horse markets. It is important to continue this historical principle of a division of the functions of city squares.

During recent decades, the vast majority of the city streets have been changed from concrete paving to the original granite paving in accordance with a plan that maps out the historical street system. In addition to this renovation and restoration of the original paving, new squares continue to be established on the basis of historical features regarding scale and materials, thus continuing the development of the historical traditions of Ribe. In this regard, the debate continues on the dilemma regarding the use of granite materials and their user-friendliness with respect to pedestrians, cyclists and the disabled. Based on this, work is being carried out to install level traffic tracks in the granite paving within a number of designated central streets (Fig. 4.5).

As part of the urban space, 'street furniture' (such as signs, waste baskets, information boards, etc.) contributes to the appearance of urban quality. The standards normally used have been developed exclusively with functionality in mind, but are of poor quality in terms of aesthetics and design, and are in no way adapted to the historical environment of Ribe, nor to the traditional materials evident elsewhere in the city. For this

Fig. 4.5 Street in Ribe – the granite floors are of vital importance to the quality of the experience of the street.

reason, in Ribe it was necessary to develop new designs for urban features and 'street furniture', which take the original city 'street furniture' as the starting point with regard to dimensions, materials, shape and details.

At the same time, the intention behind the development of this concept was to send a clear signal to the effect that the products are based on a contemporary concept of good and functional design. The concept was developed in collaboration between the local authorities, an external designer and a local producer (Fig. 4.6).

Management and regeneration action

Regeneration of the historic environment via conservation-led strategies

The strategy forming the basis for preservation in Ribe takes two sets of planning strategies as its point of departure. One set of planning strategies takes the form of a series of plans registering and appraising houses and environments worth preserving, as well as a series of physical restraints through formal protection ('listing') of individual houses. These elements can be described as defensive management tools in relation to preservation values. The other set is more active, with the objective of initiating and promoting the rebuilding and improvement of historical environments.

Preservation strategies

The 1963 *Preservation Declaration* registered on all houses within the city centre means that the local authority can require that the external appearance of a building be preserved. Any changes must have the previous agreement of the local authority. With a view to further and more comprehensive protection of the historical value of houses, a series of buildings in Ribe have been listed. These listings made by the NFNA do not confine them-

Fig. 4.6 Sign in the new urban design – the new urban design is adapted to the city environment in accordance with a new aesthetic approach.

selves to addressing external appearances; they also ensure protection of the interior parts of buildings such as floor plans and partitions, materials, etc. These two elements have been and continue to be the most important foundation for preservation efforts.

Despite these measures, further action was taken to counter any deterioration of the houses of Ribe through the formulation of a Preservation Plan for Ribe which was prepared in 1969 (Engquist, 1969). The objective of this Preservation Plan was to provide assistance and support to the public and the authorities in their work on buildings worthy of preservation. The plan describes and evaluates a substantial number of the characteristic and notable buildings of the city, but even more importantly, it also addresses a series of less distinctive, yet significant, houses.

As was mentioned above, the Ribe Preservation Atlas (1990) is the most recent registration document. This atlas comprises only part of the maps and registrations that have been prepared though field-work and analysis, and thus the atlas becomes a summary of the most important results. At its core, the atlas is a registration document that maps the buildings and environments regarded as being 'worthy of preservation'. However, due to the fact that the atlas is also capable of contrasting and comparing strengths and weaknesses, it also contributes to a debate on quality in preservation.

In terms of planning, the atlas can constitute an argument – also in political terms – for beginning the development of our contemporary history on the basis of familiar elements, yet seen from a new perspective.

The active strategies

The instruments for these strategies are more forward-looking and 'active' plans. As an extension of work on the Preservation Atlas, an *Urban Renewal*

Plan has been prepared for Ribe City, which partly analyses the urban renewal requirements of individual quarters, and partly presents an action plan for future work. The criteria addressed in the Urban Renewal Plan include the constructional standard of floor areas, sanitary/plumbing conditions of dwellings, and the nature and size of open areas within blocks.

On the basis of a summary of the individual criteria, the city's blocks are divided into three categories in accordance with the need for urban renewal. In addition to this, the Urban Renewal Plan includes summarized financial estimates for individual housing blocks, prepared on the basis of a per-square-metre assessment of the constructional standard of the floor area. Thus, there is a solid basis for renewing the historical environment. Unfortunately, the underlying legislation is often unwieldy, complicated, and demanding to work with. For instance, it does not extend in scope to business properties, a fact that severely limits the opportunities for using such properties. Thus, there are no opportunities for granting financial support to alleviate the great preservation problems related to business properties, and consequently such properties often end in a state of inertia, where neither private individuals nor local authorities are capable of completing the tasks required (Fig. 4.7).

In Ribe, the strategy for urban renewal works on the basis that funding is provided by the state to the local authority. This money is directed towards the properties and groups of properties that are of the greatest interest in terms of preservation, and where it is felt that it would not be feasible for private developers to restore them without public support. Thus, the local authority has elected to focus its attention on few, financially demanding buildings rather than on more widespread efforts throughout the town.

Fig. 4.7 The old grocer's shop has been re-erected as an urban renewal project.

The aforementioned Ribe Preservation Atlas also developed into another active strategy, insofar as it provided the basis for setting up a steering group consisting of business people from outside of Ribe and selected civil servants from the local authority of Ribe. This steering group developed a series of objectives for a new urban concept for Ribe, and three working groups prepared proposals for development of the city and its commerce, while taking into account the desire to preserve, and possibly enhance, utilization of the qualities of the city. The three main topics for the working groups were:

- development and renovation of existing tourist attractions;
- development of business activities;
- development of new opportunities for supporting tourism, and creating new activities and accommodation for tourists.

The working groups comprised non-resident experts, local authority civil servants and local enthusiasts. The work resulted in a total of twelve reports being presented in 1991, each one consisting of a specific project with descriptions of ideas, target groups, financing, organization and implementation. A series of the projects executed in Ribe during recent years, some of which have been described in this chapter, were based on these reports. Thus, the concept and method have been successful, which is largely due to the special composition of both the steering group and the working groups.

The most recent plan to direct attention to an active strategy for preservation work is the Ribe Urban Plan (described above), which addresses current preservation issues in detail and identifies/formulates the areas which require attention. One of the most significant objectives of this work is to take the historical basis as a point of departure for further development of holistic preservation and urban renewal by considering the use of funds from private, public (urban renewal) and local authority sources. These will be directed at works, operations and providing urban street furniture in a wider, prioritized context. This means attention will be directed to both improvements based on private funding, and to public expenditure on urban renewal activities which will be targeted to the weakest areas of the city. For sites where no basis exists for expenditure from public urban renewal funding sources, other appropriate solutions will be created in collaboration with owners instead.

The role of national and local authorities and agencies

Co-operation between local and national authorities is an important prerequisite for the successful improvement and restoration of historical environments. On the basis of local conditions in terms of history and preservation, local authorities are responsible for creating the planning basis required for urban preservation. This applies both in relation to the mapping and registration documents required and to the forward-looking action plans. However, during the implementation stage, where processes are initiated and fundraising is required, national authorities play a major part. Thus, the NFNA is the authorizing agency with respect to both indoor and outdoor building works on *listed buildings*. Similarly, new listings can only occur through the NFNA.

In connection with more substantial rebuilding projects and reconstruction/restoration projects on listed buildings, the NFNA may grant financial support for surveys, professional assistance and specific tasks where such support is deemed to be of special significance to the listed building. In addition to this, certain opportunities exist for making tax deductions for renovation and operation costs.

The state also plays an important role with respect to urban renewal, insofar as the Ministry of Housing and Building distributes public funding at national level, approves projects under current finance regulations, and monitors that completed accounts are in accordance with approved financial frameworks. Local authorities, in turn, may engage private consultants including a number of approved urban renewal companies. This approach has been used in Ribe. From the point of view of local authorities, state support for urban renewals is one of the most important aspects making it possible to carry out preservation work (Ministry of Housing and Building/National Building and Housing Agency, 1992).

The funding made available by national authorities is intended for preservation work throughout a wide cross-section of Danish cities. This means that very often, no real financial opportunities exist for realizing the special and very important projects arising in connection with historical cities.

To illustrate this problem it can be mentioned that on the western edge of Ribe's distinctive city frontage an old factory dominates the area. The site and appearance of this factory are not in keeping with the qualities associated with the historical city centre, and a long-term strategy involves having this enterprise moved to the industrial area and recreating a city front in accordance with the history of the city. The part of the factory which involves the least investment has now been moved outside of the city. A relocation of the remaining part of the factory would require financial investment which is not feasible in relation to the finances of the factory, and cannot be undertaken by the local authority or other investors. Even though Ribe is Denmark's oldest and best preserved city, and consequently of national as well as international importance, no realistic financial opportunities exist for solving a restoration task of this magnitude and importance. Work is currently being undertaken to examine whether funding from European Union sources might bring about a solution.

The holiday accommodation constructed according to high architecture and materials standards (Ribe Holiday Town – see above), has been erected in the same area of the city. These buildings constitute a very clear demarcation of the city against the surrounding countryside, and thus help to enhance its historical qualities. Investment in this type of venture is far more interesting to investors: hence 75 per cent of these buildings were funded by the 'Labour Market Holiday Foundation'.

There are similar examples of the inappropriate location of industrial buildings, for example, a soft-drinks factory located on the boundary between the medieval city and the surrounding meadow areas and a tractor enterprise located at the northern entrance to the city.

As responsibility for preservation work lies with local authorities to such a large extent, preservation efforts vary from local authority to local authority, depending on the political priority given to this work.

Archaeological interests

As is the case in many other historical centres, building work and road renovation in Ribe must be considered in relation to archaeological remains as the town has approximately three metres of underlying culture layers. Consequently, there is always close co-operation between the local authority and the museum *The Antiquarian Collection* at an early stage. In many cases, it is sufficient for archaeologists to carefully monitor the work – with opportunities for intervention – but in certain cases, excavation work is so interesting in view of national historical interests that actual excavations should be carried out at the site. These processes are often prolonged and costly. The *National Museum of Denmark* is responsible for large excavations, and may elect to suspend building work for one year with a view to undertaking archaeological surveys and excavation when the archaeological remains are thought to be of sufficient importance.

This can present local authorities with financial problems. If the excavation sites are privately owned, excavations are paid for by the National Museum of Denmark, i.e. by the national authority. However, if the local authority is the owner and acts as a developer, it must pay all excavation costs and fund the subsequent processing of finds. One example of this would be work in connection with an extension of a local authority care centre in Ribe, where an old abbey church was excavated. This excavation entailed further costs for the local authority to an amount of several million Danish kroner, and was indeed a valuable excavation from a national point of view. However, seen from the point of view of the local authority, such work constitutes an unreasonable drain on local funds.

The establishment of management agencies

Local authorities constitute the central element with respect to management and co-ordination of restoration of historical environments. Thus, the local authorities themselves decide on the scope and methods of preservation. The work is being supported by NFNA, but the local authorities have the responsibility. On this basis it is only rarely that national authorities or private institutions play any direct part in connection with the preparation of overall preservation policies in the municipalities.

The fact that responsibility for preservation work is placed with local authorities to such an extent means that preservation efforts are not the same in Denmark's municipalities depending on the political priority that preservation is given.

The use of guidelines for preservation work, which are drawn up by local authorities, are an example of this. Many local authorities base their preservation work on these guidelines when dealing with old urban centres, but just how these guidelines are interpreted with regard to content and attitude varies. Windows, which play a very important part in giving a building its character, are a good example to illustrate this problem. In Ribe, only wooden window frames with no aluminium edges are permitted, and for sash windows only coupled frames fitted with putty rebates are permitted. Similarly, strict requirements are made on the design, shape and dimensions of windows. In many other urban areas worthy of preservation, old windows are preferred, but the use of other materials such as aluminium and plastic, double glazing and fake sashes is permitted. Similar problems are seen in connection with other building details, shop signs, road paving, squares and so on.

When implementing urban renewal projects, local authorities often involve an existing private urban renewal enterprise which will prepare the required plans and calculations in close collaboration with local authorities and ensure that urban renewal will be carried out in accordance with legal provisions. Moreover, this enterprise is responsible for ensuring that financial frameworks are prepared and work is completed under these budgets.

The private urban renewal enterprises involved in Ribe have been the 'Urban-renewal Enterprise Denmark' and 'The Craft's Urban-renewal Enterprise' (1991).

The relationship of funding agencies

As mentioned above, public funding is mainly associated with restoration projects and urban renewal projects. The NFNA can offer financial and technical support in matters concerning restoration projects, but only in cases where listed buildings are concerned. Correspondingly, a number of tax concessions in legislation allow for tax deduction of preservation expenditure on listed buildings.

Buildings that are not *listed*, but are nevertheless judged to be *worthy of preservation*, can benefit from urban renewal projects. Urban renewal falls under the Ministry of Housing and Building which earmarks a set amount for national preservation work in the annual Finance Act. All local authorities are allotted a share of this amount in accordance with an objective basis for distribution. Local authorities can only approve urban renewal projects within the cost limits of the authority's share of the amount earmarked for preservation work.

In Ribe the strategy for urban renewal funds has been to use them on those buildings or groups of buildings which are deemed most important to preserve, and on buildings which are in such a poor condition that it is improbable that a private freeholder will be able to finance such costs without any public funding. Thus, local authorities have chosen to focus on buildings or blocks that are costly to restore, and are located in the city centre, instead of spreading their efforts throughout the city. Two building blocks, which are centrally located near the cathedral, are an example of an urban renewal project in Ribe where extensive problems with run-down and poorly adapted buildings have been solved. A commercial building was demolished to create an open space near the city's sixteenth-century city hall.

What used to be an old spinning mill was restored and converted into modern homes, and finally the very dark and narrow yards were opened up.

Alternative opportunities for raising capital are mainly found among private foundations where the boards of directors can grant funds to types of projects that fall within the scope of the foundation. Foundations have been involved in a number of projects in Ribe; one example would be the Ribe Holiday Town which was financially supported by the Labour Market Holiday Foundation.

The cost of construction of the building was about DKK 75 million, and the fund contributed DKK 52 million, making this the largest project and fund contribution in Ribe. The remaining amount was sponsored by the local authorities in Ribe, ordinary private loans and other contributors.

Previous experience with foundations as fund-raising agencies show that a prerequisite for their support will usually be that others – such as local authorities or private associations – contribute substantial monetary amounts in relation to the scale of the projects. Where these conditions are met, experience shows that foundations normally wish to have local authorities manage project content and execution. The areas where foundations wish to exert influence involve project design and architecture. This means that foundations may appoint architects or may be on the panel of judges for the architectural competition. Moreover, experience shows that foundations place great emphasis on keeping projects in accordance with local authority guidelines and policies. On this basis, projects are usually carried out in a positive dialogue between foundations, users and authorities.

In connection with the project *Ribe Byferie* (Ribe Holiday Town), of which the fund contributed a rather substantial share of the financing, the fund's board and architect became more involved in the project's content than usual, just as the architect was involved throughout the entire period of the building's completion.

The need and opportunities for flexibility to adapt historical buildings to new uses

The vast majority of the buildings being repaired in central Ribe are used for dwellings and retailing in connection with shopping streets. The NFNA provides authorizations with respect to building works on listed buildings. In its decisions, the Agency accords great importance to the fact that these works should take into account the values for which the buildings were listed, that the building should have an appropriate function and be maintained in the long term. As a consequence, opportunities for using buildings for completely new purposes can, in certain cases, be limited.

Another building category is former business buildings which, by virtue of their architecture and materials, are significant elements of the historical environment and are deemed *worthy of preservation*. It is important that opportunities exist for putting these buildings to new uses, possibly in connection with the erection of new buildings or annexes.

One example of this would be the museum *Vikings of Ribe*, which has been accommodated within a former Electricity Building dating from 1924 and regarded as being very characteristic of its period. This building forms part of the total building complex of the museum and sets the standard for the proportions and design of the new buildings – naturally with a contemporary aesthetic approach (Fig. 4.8). Another example is the conversion of a former railway 'round-house' into a culture centre, where the historical railway building is even more clearly separate from the extensions subsequently erected. The permitted scheme required that it would still be possible to view the very distinctive 'round-house' as an independent building with intact architectural design and proportions. In addition, the new buildings were required to adapt themselves to this imposing building in terms of architecture and design, while still adding a new historical layer to the city in a harmonious fashion. Generally speaking, this approach underpins all projects where old historical buildings are incorporated into new contexts, physical and functional.

Environmental management

Management of traffic and transport within the historical environment

Total traffic in Denmark rose by approximately 70 per cent during the period 1980–96 and will continue to rise. This increase in traffic is also clearly felt in Ribe, both on the main roads in and around the city and in terms of local traffic inside the city. The traffic structure in conjunction with the growth in traffic in Ribe presents various problems for residents, the many visitors, and the physical environment of the city. In particular, the city receives approximately one million visitors each year, and this, in conjunction with the use of the old city centre for dwellings and retailing, means that the medieval streets are subjected to tremendous stresses. Large tour buses and lorries constitute a particular nuisance in the narrow city streets. At the same

Fig. 4.8 The museum Vikings of Ribe – the former electricity building, with its extension, is an example of a new use for an old building.

time, the narrow streets mean that only few parking spaces are available within the city centre; thus parking occurs in the narrow, medieval streets.

These problems are exacerbated by the fact that the buildings of the city centre are situated on deep, old cultural layers and do not have the foundations required to withstand the tremors and vibrations caused by heavy traffic. The result is that the cobblestone streets and culture layers continue to be destroyed, and hence the old houses and urban environment as such are also affected.

During recent years, the local authority of Ribe has addressed these issues in the Urban Plan. The objectives of the plan include statements to the effect that infrastructure work will be carried out in order to create easy accessibility between individual city areas and the general road structure. It is also stated that the medieval city centre should be protected from traffic. The opportunities and tools to be used for meeting these objectives can be found at various levels:

- Work is being carried out in co-operation with authorities on moving the present main road further towards the west. This will mean that the present ring road, in conjunction with a new road to the east of the city centre, will be able to function as an inner ring around the historical city centre, thus distributing traffic to individual destinations without traversing the historical city centre.
- The establishment of additional large parking areas along the ring road will mean that tourists and shoppers from the surrounding area will be able to park in these spaces and cover the relatively short distance to the city centre on foot or on one of the city buses. At the same time this will allow for parking in the old part of the city to be reserved mainly for inhabitants.
- The main street of the city is now a pedestrian area. During recent years great efforts have been made to have all the old streets converted into a form of pedestrian area where all traffic must be carried out on pedestrians' terms, and where motorized traffic is allowed only at very low speed. As a supplement, a system of restricted-access streets has been established to cater for internal, local traffic.

In the event that these measures and tools should prove insufficient, more radical measures may be considered. These could include signs posted at access roads to the city centre, banning all access, or banning all access to the city centre except for residents and shopkeepers. This solution has been successfully used in other Danish cities.

In order to create a balance between the functions of traffic areas, in terms of handling traffic and in terms of their importance as significant urban elements, it is considered vital to have high quality designs of streets and squares. These constitute an important factor in the experience of the city environment.

The special feature of the work on problems regarding traffic policies is that no clear and unambiguous solutions exist which address the total set of problems. As a consequence, a series of consumer interests will be set aside when assigning political priorities to individual solutions. A lack of political resolve to make the necessary decisions and assign priorities as required will mean that no action can be taken to follow up objectives, and that no lasting and general solutions can be created.

Tourism and heritage management

Every year, Ribe welcomes a massive influx of visitors whose main objective is to experience the well-preserved historical city centre and its special atmosphere. A large proportion of these tourists are one-day tourists from the large holiday-house districts along the west coast of Jutland. The city has great potential in terms of tourism, but a corresponding commercial effect has failed to emerge, since one-day tourists do not provide the type of business that is significant for city trade. Surveys have indicated that only about 10 per cent of the business turnover within Ribe is actually derived from tourism and tourist activity.

A report on retail sales from 1995 indicates that 15 per cent of turnover stems from customers who do not live in the Ribe area (Planning enterprise Erik Agergård, 1995). This 15 per cent is thought to be equally distributed between tourists and other groups. Turnover generated by tourists comprises 7–8 per cent of the total turnover, which corresponds to approximately DKK 40 million and covers purchases made by tourists. In addition to this, there are other and most probably larger sums that tourists spend in hotels, restaurants, etc., which generate far more income than the purchase of daily groceries. Thus a tourist brochure from 1991 indicates that the total turnover generated by tourism is thought to be about DKK 140 million. However, these figures may be subject to some uncertainty. Considering the recent tourist initiatives, for example, *Ribe Byferie* (Ribe Holiday Town), Ribe's Viking Museum, and the *Viking Centre in Lustrupholm*, the total turnover must be increasing.

As is the case in other tourist cities, some of the businesses in the town are aimed at tourists, such as souvenir shops and ice-cream booths, cafés and restaurants. A characteristic feature of some of these shops has been that they vary greatly in terms of quality and originality, and consequently this segment of commercial life in Ribe has been unable to match the quality reflected elsewhere in the city, thus failing to make full use of tourism potential. At the same time, there has been a tendency among certain shops aimed at tourists to open during the tourist season only.

The number of visitors now coming to Ribe means that saturation point has been reached, and an increase in this number would occasion a negative shift in the balance between the need for tourism and local acceptance of tourists. As a consequence, tourism strategies in recent years have not attempted to increase the number of tourists, but have, instead, aimed at encouraging tourists to stay longer and at extending the season with a view to distributing pressure over a longer period.

Such tourism management is very difficult to accomplish, but it is possible to create a 'product' which is particularly attractive to specific target groups.

In Ribe, initiatives have been directed towards tourists demanding experience and authenticity in an historical environment. By expanding accommodation facilities, the necessary basis for this type of tourism has been created. At the same time, work is being carried out on changing the business concept to improve quality, and there is great pressure to ensure that no more seasonal shops emerge.

The cultural role of the historical centre

The development of and attention to tourism in Ribe during the last 5 years has a clear basis in the historical and cultural environment of Ribe. Substantial investments have been directed at activities that have helped to document the city history, to communicate knowledge, to assist in strengthening Ribe's cultural role in relation to tourism, and especially in relation to local residents as it is vital that the initiatives adopted are attractive to this group.

The most important recent projects carried out include the museum Vikings of Ribe, which relates the Viking and medieval history of the city in an inspiring, contemporary manner and the Lustrupholm Viking Centre. This centre brings Viking history to life by recreating a working Viking settlement with markets, a chieftain's hall, stables with livestock and workshops with working artisans and craftsmen. Moreover, a *Wadden Sea Museum* has been erected a few kilometres south of Ribe. This museum communicates knowledge about the completely unique and unusual flora and fauna of the entire Wadden Sea. In addition to this, a series of cultural events, such as *Cultural Nights*, are held throughout the year and help keep the city vibrant and alive.

It is important that historical centres consciously take on responsibility for communicating and relating their historical heritage in a manner based on concepts of authenticity and quality, and that they contribute to the general understanding and appreciation of our national history. This means that these concepts must be formulated clearly and that there must be a resolve to reject projects and initiatives that do not meet these requirements. The difficult balancing act consists of having the city work both as an historical centre, and as the dwelling and workplace for citizens in a modern society. In this way, the city remains a 'living city' rather than a 'museum town'.

Sustainability

If sustainable development of historical environments is defined as development that creates new, modern historical layers without compromising the future, then a good basis has been created for the upkeep and preservation of historical towns and cities. It is important to realize that historical towns and cities were created throughout a thousand-year period, where developments have occurred in very small steps. With the technology and knowledge available today, it is possible to change and destroy irreplaceable values within a few years, if we do not display continual awareness, assessing the consequences of our actions in a historical context. This is not to say that change should be prevented, or that everything should be caught in a historical freeze-frame. Quite the opposite, it is a question of managing the change; history is a chain of changes; new things arise and the old disappears – indeed, this is why the old city centre of Ribe is situated on a culture layer of a depth of three metres.

Change in itself does not present a danger; however, the rate and frequency of change does. Too many changes happening too quickly may upset historical continuity. The trend has been for those tools of preservation and planning used by us in the spirit of preservation to take the shape of static plans and registrations, which do indeed constitute important bases for preservation, but which also reflect a fear of creating new history ourselves.

Currently, there are no useful tools targeted at enabling local authorities to create new developments with a basis in history. Consequently, we all too often find that changes are made random and hence lack quality, consistency, coherence and sustainability.

A comparison of the most important historical cities in Denmark reveals a considerable difference in the level and quality of preservation work, and also in the interpretation of the term sustainability. In Ribe great thought has been given to the question of sustainability in the development and preservation projects that have been carried out. This has ensured that projects such as *Ribe Byferie* (Ribe Holiday Town), the Viking Centre, the urban renewal project near the cathedral, the new homes constructed in Slotsgade and the rebuilding of the railway round-house have contributed to supporting and enhancing the character of the city, both in content and form. Each project has been assessed and adapted so as to complement the historical context and the city itself.

A future task will be to solve the immense traffic problems the city has, and create sustainable solutions. One of the means to do this is the Ribe Urban Plan, which takes a holistic approach to sustainability and assesses the context in which measures will be carried out. This should also be a central element in future plans.

Today, state listings protect 'elite' historical buildings. However, maintaining and creating sustainable historical urban environments requires that broader and more active management systems should be developed, focusing on both preservation and development seen from the perspective of some clearly-defined quality requirements. The closest we have come to such a tool is the aforementioned Preservation Atlas, which lists the city's assets in a new and more cohesive manner; however it does not include strategy and action plans.

Those local authorities that have continued working with the registration and assessments afforded by the atlas have done so in the form of preservation guidelines and local preservation plans. These tools can also be described as passive tools, and are only taken into use as regulatory documents when citizens wish to change existing conditions. There is a discrepancy between the Preservation Atlas and the local preservation plans that should be rectified by active measures that pave the way for action. Two previously mentioned measures have been applied to rectify this discrepancy in Ribe. The measures in question are those development projects which were initiated in 1991 as an extension of the atlas work, and the Ribe Urban Plan, which the local authorities prepared in 1999, and which particularly focuses on objectives, areas for initiatives and projects.

We can offer no solution here as to how the process from atlas to objectives, and areas for initiatives to binding plans and preservation documents can be developed into a useable tool and thus contribute to sustainable development.

Also the need for financial sustainability must be mentioned. Thus, there is great need for innovation with respect to more general and high-level work to secure Denmark's most important historical environments where local authorities cannot carry out this task themselves. At present, such preservation and care for our cultural and historical heritage is largely left to individual local authorities to manage.

In Ribe, there are four important projects and problems that the local authorities cannot solve on their own. These are the renovation of the cathedral square, the renovation and improvement of the city's harbour front, and relocation of the aforementioned industrial enterprises in the southern part of the city and near the northern entrance. Attempts are still being made to procure financial backing within European Union funding or as part of the DKK 90 million the state has earmarked for such projects at a national level. At present it seems improbable that the necessary means will be found through these channels.

Conclusions

Previous generations have bequeathed upon us a unique and singular historical environment in Ribe city centre, where the traces of history can easily be followed back in time for several hundred years. When strolling around the city, experiencing the houses, streets and breadth of history, you can almost feel the silver buttons sprouting from your waistcoat and hear the clang of iron cartwheels against cobblestones – despite the fact that Ribe is now buzzing with modern people and cars.

At the same time, the gift we have been given places a heavy burden of responsibility on the public of our time. Despite its many qualities and unblemished environments, Ribe is also a city full of paradoxes and conflicting interests, torn between past and present. If we are to be able to handle this inheritance, we will require diligence, careful thought and foresight. This is not just an issue for democratically elected decision-makers and civil servants at national and local levels, it is also relevant to everyone interacting with the city and thus influencing its development each day.

Postscript

So far, so good for Ribe, not only one of the oldest cities in Denmark, but also one of the Danish cities that has achieved most with regard to preservation work. This is why the National Forest and Nature Agency chose

Ribe to be represented, from a local point of view, in the examination of the Management of European Historic Centres. However, it is worth mentioning that there are other tools in urban and building preservation than those that local authorities have chosen to work with in Ribe.

A series of legislative amendments in 1997 enabled local authorities not only to identify which buildings were *worthy of preservation* by way of a local plan, but also to identify these buildings in the local authority plan. At the same time it became possible to appoint committees dedicated to improving buildings and thus subsidize construction work on buildings worthy of preservation. For example, this could mean helping the owner of a building worthy of preservation to have a duplicate of the original, but now rather run-down front door made, instead of replacing the door with a standard product.

In addition to these more formal tools, there is a range of less formal, but none the less important measures. These include the many opportunities for creating local interest in preservation work, for example in preservation associations and other local preservation organizations. If local authorities continue to keep the public informed about local preservation values, it may help to enhance the interest of the public. The NFNA's Preservation Atlas is a good place to start, for example, by organizing city walks and other events concerning building preservation.

To support this work, alongside the aforementioned legislative amendments from 1997, it became possible to appoint regional cultural environment councils, which subsequently also happened in other parts of the country. The primary task for these cultural environment councils is to contribute to an open and informed debate. The main function of the councils is one of technical insight, whilst decisions are still the responsibility of the individual city councils.

Moreover, the NFNA is very aware of the fact that local authorities may need advice and guidance at many levels. Guidelines regarding the use of local plans for preservation, information pamphlets about the preservation of buildings and projects in which the NFNA is currently involved, in co-operation with the local authorities, are a few amongst many good tools that local authorities can utilize.

As mentioned several times in this chapter, Danish local authorities are responsible for preserving non-listed buildings that are nevertheless worthy of preservation. This task requires a policy concerning the area, an active use of the various planning tools, as well as continued information about the qualities of buildings worthy of preservation – and how to take care of these buildings. The NFNA is following developments with great interest and hopes that Danish local authorities will take advantage of the assistance that the agency can offer them.

References

Craft's Urban-renewal Enterprise (1991): *Urban renewal study in Ribe.*
Engquist, H.H. (1969): *Preservation Plan for Ribe.*
Ministry of Environment (1990): *Preservation Atlas Ribe.*

Ministry of Environment and Energy/National Forest and Nature Agency (1995): *InterSAVE, International Survey of Architectural Values.*
Ministry of Housing and Building/National Building and Housing Agency (1992): *Urban Renewal and Housing Improvement in Denmark.*
Planning enterprise Erik Agergård (1995): *Retail Sales in Ribe 1995–2000*
Ribe City Council (1963): *Declaration of preservation of 16 April 1963 of 550 properties in the inner city.*
Ribe Municipality (1997): *Good Advice on old Houses.*
Ribe Municipality (1998): *Local Authority Plan for Ribe 1998–2009.*

Other sources of information:

Bencard, M. (1978): *Ribe over the last 1000 years.*
University of Southern Denmark (1991): *Regional research: Memorandum 23/91 (1991): Tourism in Ribe.*

Bruno Coussy

Rochefort, France

Introduction

The management of historic towns through planning and the conservation of the built heritage are concerned with very particular notions of time. On the one hand, planning is above all a precise science, defining objectives to be achieved and therefore situated in the future. On the other hand, while 'heritage' refers to the notion of a legacy, it is essentially the 'shared memory' and therefore belongs to the past. Putting these two notions side by side thus takes us straight to the essential questions of the existence of any community: 'Where do we come from?' 'Where are we going?' 'Who are we?'

From this standpoint, the way in which the community formulates land-use and development plans is of particular importance. Will it do this on the basis of normative, quantitative criteria, for example: on the basis of an expression of the needs of a population of so many inhabitants – criteria that are reassuring by virtue of their universal nature. Or should it be on the basis of more sensitive criteria connected with the specific characteristics of the place and its inhabitants. In this case the criteria are more difficult to determine because they are connected with the exceptional nature of each town, as regards both its built heritage and its social make-up.

The example of the town of Rochefort, in France, seems to be particularly interesting in this respect. During the past 20 years, thanks to a very pro-active heritage policy, it has passed from the status of 'Garrison town to be avoided' to that of 'Exceptional 17th century naval dockyard town' (Coussy, 1988).

The town of Rochefort has applied, some years ahead of time and with a great deal of intuition and pragmatism, the ideas contained in the *International Charter for the Conservation of Historic Towns and Urban Areas*

(ICOMOS, 1987). In particular, this can be noted in the following principles of the Charter:

1 In order to be most effective, the conservation of historic towns and other historic urban areas should be an integral part of coherent policies of economic and social development and of urban and regional planning at every level.
2 Qualities to be preserved include the historic character of the town or urban area and all those material and spiritual elements that express this character.

. . .

4 Conservation in an historic town or urban area demands prudence, a systematic approach and discipline. Rigidity should be avoided since individual cases may present specific problems.

In fact it can be noted that, beyond the historic centre, it is a matter of knowing what attitude should be taken with respect to the town. Hence it is a way of living together, and being aware that the forms of conservation of the heritage have implications that are much more political than technical.

The planning and policy framework

Historical background

Rochefort is a small town of 30,000 inhabitants situated on the Atlantic coast half-way between the major urban centres of Bordeaux and Nantes.

In 1970, its situation was particularly disastrous. The inhabitants deserted the old centre, preferring to build comfortable new dwellings on the periphery of the town. This was the movement of a population that drew no pride from its history and regretted that the town did not resemble its neighbour, La Rochelle. In fact, Rochefort had lost its identity.

And yet Rochefort has an exceptional history. In 1666, Louis XIV, who did not have a large fleet compared with the English and Dutch, instructed Colbert to find a site for a naval dockyard on the Atlantic coast. He discovered the estuary of the River Charente, protected by the islands of Ré, Oléron and Aix and, some 20 km upstream, a vast area of marshland. It was here, in a loop of the Charente, that the town of Rochefort was to be created (Fig. 5.1). Intense activity soon developed along the Charente and by 1700 the town had over 10,000 inhabitants. For 250 years, Rochefort lived to the rhythm of the central government's naval policy, but in 1926, now considered too remote, the Rochefort dockyard was no longer economic. Its closure cost 6,000 jobs and triggered the decline of the town.

Left aside by the post-war economic boom, Rochefort continued to stagnate. But the heritage of the naval dockyard, and of the town, deteriorated very little and in 1970 Rochefort still had remarkable monuments and, above all, a remarkably homogenous urban fabric (Figs. 5.2 and 5.3), both in the town centre and in the surrounding districts.

The political grouping that won the local elections at that time was made up of young entrepreneurs who believed in the renewal of their town and

Fig. 5.1 Rochefort – an island in the middle of marshes along the River Charente, a former naval dockyard. (Photograph: A. Barathieu and B. Coussy.)

called itself *Rochefort's Renaissance*. This name itself immediately suggests another view of the heritage. It does not speak of the development or extension of the town, but of renaissance or 'rebirth', a new departure on the basis of other values, another relationship to history.

At that time, the French state was trying to enhance the status of small towns and launched a procedure known as *Contrat Ville Moyenne* (Medium Town Contract), endowed with specific funding. Rochefort was to benefit from this contract with the central government.

To carry out the necessary preliminary studies, the elected representatives contacted a town-planning college in Paris (*Institut d'Etudes Politiques – Cycle Supérieur d'Urbanisme*) and a number of students came to analyse the town. The approach is interesting, because this college follows a multi-

Fig. 5.2 A remarkably homogeneous ancient centre. (Photograph: A. Barathieu and B. Coussy.)

Fig. 5.3 A simple and repetitive architecture. (Photograph: A. Barathieu and B. Coussy.)

disciplinary approach to the problems of town planning and development and attracts architects, sociologists, lawyers, etc. These students were impressed by the immense heritage, essentially military, capable of being re-used, and by the grandeur and unity of the old town centre. The main proposals were as follows.

- All of the operations should be located in the town centre and anything that helps rehabilitate the centre is good for the town as a whole.
- Re-use of the built heritage is preferable to new construction. New building should be permitted only after it has been established that no existing building can be used.
- The above two proposals aim to reconstitute and take advantage of the naval tradition acquired by the town in the course of its history, even though the shipbuilding industry has now disappeared.

These three proposals were to guide all urban development policy over the following 20 years.

It is interesting to note that the local council again made same request for help to the Institut d'Etudes Politiques at the close of the 1990s, in order to renew the thinking about the town and draw up a medium- and long-term action plan.

Rehabilitation policy and action

The decision to rehabilitate the *Corderie Royale* (Royal Ropeworks) in 1975, and to create a park around it, was the keynote operation of this revival of the heritage. The old rope factory building, 400 metres long, dates from the end of the seventeenth century and is the feature most evocative of the

Fig. 5.4 La Corderie Royale (Royal Ropeworks) before restoration in 1960. (Photograph: A. Barathieu and B. Coussy.)

dockyard (Fig. 5.4). Its complete rehabilitation (only the walls remained), and the different uses to which it was put (National Coastal Conservatory (CNL), French League for the Protection of Birds (LPO), Chamber of Commerce and Industry (CCI), municipal library and media centre) was the starting point for the renewal of the town. Through a series of phases the work was carried out between 1976 and 1988). The completed project became a symbol of renewed pride.

The creation of the park around the Royal Ropeworks was to be an important factor for the better understanding and enhancement of the town. When a competition was held for the design of this park, whilst most of the candidates proposed a project linking the town and Ropeworks, just one candidate stood out by opening the orientation of the entire site and the building not towards the town, but towards the river. He called his project *Le Jardin des Retours* (Garden of Homecomings), stressing the return of all the expeditions that brought home to Rochefort plants discovered in distant countries. The town backs up to the dockyard, which is set on the river where ships were carried out to the open sea. The main method of access to Rochefort was by sea, and to make it a purely terrestrial town would be to deny its naval past (Fig. 5.5). In its turn, this new vision of the town, proposed by Bernard Lassus, landscape architect, was to have a profound influence on subsequent planning.

More recently, the decision to rebuild *L'Hermione*, the frigate on which La Fayette sailed in 1780 to fight on the side of the American rebels, again corresponds to precise objectives. Rising above the roofs, and the only high points in the town, the ships' masts used to be vital features of the urban landscape. The mast of *L'Hermione* (some 50 metres high) will be situated at the true centre of gravity of the area formed by the dockyard and the old town centre. It will be the only high point in the town where the heights of the buildings are even and there are no bell-towers. What is more, the Rochefort

Fig. 5.5 One of the numerous forts situated in the estuary responsible for defending the Naval Dockyard (L'Arsenal de Rochefort). (Photograph: A. Barathieu and B. Coussy.)

Naval Dockyard (*L'Arsenal de Rochefort*) can hardly be understood as such without the presence of something representing its function: building warships for the King of France. The shipyard is now a great success as a tourist attraction, with over 300,000 visitors in the second year of operation.

The monumental restoration of the Royal Ropeworks was accompanied by a policy aimed at using the existing built heritage, virtually exclusively, in the rest of the town. For example, a district multi-activity centre was created in a former naval apprentice school, the former barracks were converted into a Court of Justice, the powder magazine into a place for rock groups to practise, and the general dockyard stores into workshops and offices, etc. The gradual departure of the French Navy from Rochefort meant that many buildings of great architectural and historic value came on to the market. This process is continuing today with the closure of the naval air service base (300 hectares, 200,000 m^2 of floor space to turn to new uses!). This systematic rehabilitation and re-use of the heritage had a considerable incentive effect on large private enterprises (Matra aéro-spatiale, Zodiac), who installed their management services department in old dockyard buildings (Sculptures Workshop, Port Management building).

The restoration policy also came to include housing, with large-scale action by the Municipal Public Housing Office (HLM) rehabilitating some 300 dwellings in 20 years, on top of which came twice that number of private dwellings. This housing rehabilitation has been of vital importance for the old town centre, all the more so as this centre is very extensive (50 hectares) for a town the size of Rochefort (see Table 5.1 below).

Rochefort Architectural Charter, land-use planning and accompanying actions

On another level, the urban heritage – its morphology and its typology – was closely analysed in the *Charte Architecturale de Rochefort* (Rochefort Architectural Charter) (Coussy *et al.*, 1998). This document was directed towards the public to explain the concepts of urban form to be promoted concerning the existing architecture and contemporary architecture. This Architectural Charter, at first limited to the old districts, was recently extended to the new districts and to the whole of the municipal area. Defining the urban landscape in graphic fashion, it serves as a basis for the municipal land-use plan (*Plan d'Occupation des Sols* – POS) and all the rules and regulations concerning land use. The Architectural Charter followed very closely the analysis of the values to be preserved as set out in the *Charter of Historic Towns and Urban Areas* (ICOMOS, 1987): The Rochefort version reads:

> 2. The features to be preserved are the historic nature of the town and all of the physical and intangible or spiritual elements that go to make up its character, in particular:
>
> - The historic development patterns of the town, as these have emerged over time.
> - The special relationships between the different urban areas: built areas, open spaces, planted spaces.
> - The (internal and external) form and aspect of the buildings themselves, as defined by their structure, volume, style, proportions, materials, colour and ornamentation.
> - The unique relationship between the historic town and its natural and man-made surroundings.
> - The various functions that the town has acquired in the course of its history.
>
> Any threat to any of these qualities would compromise the authenticity of the historic town.

The Architectural Charter, over 8,000 copies of which have been printed, is in everyday use by everybody who is concerned with the heritage of the town.

Another important current project is the transformation of the Museum of Art and History into an Interpretation Centre for the town, bringing together the collections of the Museum (in particular a remarkable bas-relief plan of 1835), heritage workshops used to raise awareness among school children and the Tourism Office. These actions have been directed towards promoting quality tourism in Rochefort. This Centre of Interpretation is intended to relate the history of the town to the inhabitants and to initiate the discovery of the town and the naval dockyard.

Attention should also be drawn to the valuable work carried out by associations concerned with small heritage items (wells and fountains, names of streets engraved on façades, etc.) that they endeavour to protect. Their action is far from negligible in making the old districts look their best.

In parallel with the efforts made in the town centre, particular care is being taken with new developments. These include new streets in the suburban area with new semi-detached and town houses, keeping the traditional aspect of this area; houses grouped around a common courtyard on the outskirts of old farming hamlets; and a simple grid layout for large-scale housing developments on the periphery. As for the old centre, the notion of 'reference' is a constant feature, each new district making reference to the neighbouring traditional district, whilst at the same time introducing the necessary modern elements.

Management and regeneration action

As stated above, the re-use of the buildings in the old centre has been systematic over the past 20 years. Today we find very diverse uses such as a Music Conservatory in an old powder magazine, wrestling rings in a deconsecrated church, warehouses in the Royal Foundry, a Court of Justice in a former barracks, etc.

For the centre to be able to live it needs activities and these are appearing within the old town walls. A number of factors have facilitated this. First, the historic centre of Rochefort is particularly large for a town of this size (50 hectares); secondly, the streets are broad, permitting traffic circulation and parking; and lastly, the gradual departure of the Navy has released a number of buildings that have been well-maintained over the years, due to the fact that they were that occupied right up to the time they were put on the market. Another contributing factor was the choice by the Municipality to conserve the external silhouette of the buildings and, in most cases, to completely rearrange the interiors of the buildings, thus permitting new uses at acceptable cost.

To implement this policy of regeneration of the old centre, the elected representatives set up a Town Planning Service in the Municipality, very close to the Mayor and with broad powers concerning:

- the drafting of the Land Use Plan;
- the management of the major projects (Royal Ropeworks, L'Hermione, rehabilitation of the town centre dwellings, reopening of the old commercial port and development of the quays, etc.);
- examination of building permits;
- institutional communication;
- and the setting up of Heritage Workshops.

This cultural approach, necessarily being multi-disciplinary (involving town planners, architects, etc.), created the opportunity to set up a small team, which was highly motivated by the prospect of restoring the past grandeur of the town. For each operation it sought adequate financing from the territorial authorities (Department, Region), the central government and through European Funds.

As regards housing, which is an essential part of the process of revitalizing the old centre, the town benefited from several agreements through the *Opérations Programmées d'Amélioration de l'Habitat* (OPAH – Programmed Operations for the Improvement of Housing), concluded between the central government, *l'Agence Nationale pour l'Amélioration de l'Habitat* (ANAH – the National Agency for the Improvement of Housing), and the Municipality of Rochefort. The latest OPAH (1996–98) concerned the rehabilitation of vacant dwellings situated above shops, all too often used as storerooms with access from the shop. For its part, the Municipal Public Housing Office (HLM) used soft loans from the central government to rehabilitate many buildings that the Town acquired and leased back (see Table 5.1).

The decompartmentalized and transversal nature of the municipal services are certainly important factors for the good management of the old centre: there is not one person responsible for buildings, another for utility networks, a third for protection, etc. Everybody is fired by the same initial idea ('What is good for the centre is good for the town') and can suggest this or that acquisition, this or that re-use. Thus the general secretary of the town, his deputy, the architect, the special adviser and a former municipal counsellor are all graduates of the same town planning school, which helps decompartmentalized work on the basis of different sensibilities but fired

Table 5.1 City of Rochefort, housing rehabilitated through the support of the National Agency for the Improvement of Housing (ANAH) – OPAH from 1991 to 1993 and a pilot operation on rehabilitation of vacant space above shops for housing (to rent) from 1996–8

	1992	*1993*	*1994*	*1995*	*1996*	*1997*	*1998*	*Total 1992/98*
Housing rehabilitated through the support of ANAH (Owners who lease their property/landlords)								
City of Rochefort	59	60	43	23	56	29	43	313
Of which OPAH	*55*	*58*	*22*		*20*	*4*	*10*	*169*
Housing rehabilitated through programme for improving housing (PAH) (Owners living on their property)								
City of Rochefort	21	43	18	8	21	26	20	157
Of which OPAH	*20*	*42*	*17*					*79*

Source: DDE 17.
Notes: The rehabilitation operation of vacant stores located above shops (for renting) (34 properties rehabilitated between 1996 and 1998). Cost: 5,802,135 FF.
Total amount of subsidies: 1,365,817 FF, of which 1,225,604 FF is from ANAH and 138,213 FF from the City of Rochefort.

by the same will to develop the town. This internal decompartmentalization is reinforced by constantly calling in people from outside who can take a fresh look at the town.

At the start of the twenty-first century the process is taking on a whole new dimension. The fact is that while the naval dockyard and the town of Rochefort form a homogenous whole, they themselves fit into a vast defensive military complex which extends to the islands responsible for defending the estuary of the Charente. This military heritage includes numerous fortresses and small forts, powder magazines, mast-timber ponds, fountains for supplying vessels with fresh water and other installation structures within this estuary landscape, which is covered by some twenty municipalities. The procedures used to restore the heritage of Rochefort to its rightful place should therefore be gradually extended to the scale of this much larger unit.

Environmental management

In a small town like Rochefort where the area of public spaces associated with the original grid layout is very substantial, the problems of traffic movement and parking are by no means as acute as elsewhere. Nevertheless, the traffic management system is constantly under review and subject to change, due to an ever-increasing number of cars.

The main feature of the traffic management system is the systematic elimination of traffic lights in favour of mini roundabouts so as to enhance traffic flow. In addition, an active policy to encourage cycling is pursued (30 km/h zone in the historic centre, cycle lanes and dedicated cycle tracks further from the town centre). As far as parking is concerned, Rochefort has the good fortune to have very large courtyards corresponding to the old clear-fire zone before the ramparts that can be used as parking areas for visitors to the old centre. A new signposting system is being installed to oblige tourist vehicles to traverse the old town when going to visit the dockyard. Lastly, the town has created two public transport lines, using two shuttle buses each carrying about thirty passengers.

On another level, Rochefort is pursuing a very proactive policy to win land from the periurban marshland to promote agro-environmental areas and avoid the creation of wastelands. The procedure used to enhance these natural areas is identical to that used for the built areas and involves the following stages:

- Detailed and sensitive analysis of what exists;
- reconstitution of the historic evolution of the landscape and of the sources of enhancement or degradation;
- formulation of objectives;
- agricultural development;
- environmental protection;
- reconstitution of fauna and flora;
- urban recreation;
- limitation of urban development, etc.;

- search for solutions and establishment of an action plan (pastures, reconstitution of the ditches, controlled flooding, etc.);
- appropriate action.

Just as the aim of the Architectural Charter was to express the identity of the town, the aim of the study of the natural areas is to reveal the terrestrial island of Rochefort, situated a few metres above the marsh and, on the scale of the broad estuary landscapes. The concept envisages a single area, made up of islands in the sea and islands on land, located in the marsh and bearing witness to the presence of a sea now withdrawn. It is therefore now the water of the ditches and channels that reveals the landscape, which should organize it and define the limits of the town.

Tourism and heritage management

Contrary to what people sometimes think, the rehabilitation of Colbert's Dockyard and the historic centre was never done to attract tourists. It was simply a matter of restoring meaning to this town and a sense of pride to its inhabitants through restoring symbolic historic monuments (Royal Rope-works), creating an urban park (Garden of Homecomings) and renovating the buildings in the town centre. The scale here is more convivial than in the peripheral districts whose development was dictated by the automobile.

It was only gradually that the Municipality realized that tourists appreciated the atmosphere of Rochefort. Thus, through pursuing a purely municipal policy, it was still possible to interest tourists (Fig. 5.6). An article that appeared in a town planning review in 1986 was entitled: 'Rochefort was not expecting the tourists, but they came all the same' (Coussy *et al.*, 1986). The direction was clear – it was a matter of pursuing a policy for the benefit of the inhabitants and saying that if it also attracted tourists, then so much the better.

The construction of *L'Hermione* was thus immediately associated with that of the Airbus manufactured nearby to show the historic continuity between the frigates of 1780 and the aircraft of 1990, both destined to cross the oceans. Sections of Airbus lie alongside the stocks of timber for *L'Hermione* and affirm the local will to develop, rather than to sink into nostalgic addition to the past.

The Municipality recently launched a competition for the rearrangement of the Municipal Museum on the theme 'Discover your Town' combining three functions: the museographical journey, the Tourism Office and heritage workshops. The aim of the first is to let people discover the history of the town, the aim of the second is to promote the town to visitors and aim of the third is to arouse awareness among school children. There will thus be a single place housing all the resources for an initiation to Rochefort's heritage, without it being very clear whether the target public is the inhabitants or the tourists.

This serene attitude of taking account of the local needs is opposed to any kind of folklore of interest to tourists only and which in any case would only be self-defeating in the end.

Fig. 5.6 Arrival of the English frigate 'The Rose' in front of the Corderie Park. (Photograph: A. Barathieu and B. Coussy.)

Sustainability

The main risk of unsustainable development today is connected with the transformation of the town into an object of consumption by both its inhabitants and by visitors. The former pay their local taxes and expect a number of 'services' in return, whereas the latter are more interested by the local folklore than by a true meeting with the people. Both groups destroy the ideal of a united community attached to a place.

On the contrary, sustainable development finds its expression, not in immediate pleasures, but in the creation of a long-term urban project concerning all the inhabitants at different levels. The actual objective of the project is less important than the process of implementation in which all feel linked by common purpose.

The image of the oriental mandala is very expressive in this respect, where a group constructs a magnificent image with a little coloured sand and then throws the sand away when the image has hardly been completed. The process of collective working towards the same goal makes the object created unimportant. This is a world away from the consumer society and the spectator society.

Conclusions

All too often, thinking about urban development suffers from a lack of real knowledge of the place and the population. Based on normalized needs, (those of the motorist, for example, which are the same everywhere), the reasoning is limited to quantitative criteria (average housing density, for example) and simple zoning. In this approach, the monument or old district becomes a category in itself, which should be protected but which is in no way a source of inspiration. In this type of thinking it is accepted that modern man has freed himself from ties to a place.

Conversely, integrated planning implies having a very good knowledge of the spirit of a place before taking any action. The spirit of a place is not the inventory of the heritage (in the sense of a list of objects) but the sensibility with which this heritage has responded to the many demands of man over the centuries.

In the case of Rochefort, a basic examination would see only a small town of 30,000 inhabitants, the size of the peripheral marshlands, the absence of a bridge over the Charente near the town centre, and a place 'hampered' in its development by military wastelands, etc. It was in line with this 'rational' approach that a bypass was planned in 1970, on the site of the Royal Ropeworks, which would have destroyed all memory of the place and all connection between the town and its river. Similarly, when part of the dockyard was converted into an activity zone after the war, it was a matter of moving towards a modern industrial estate, gradually eliminating the old buildings, whatever part they had played in the history of the town. The public area was in fact simply a central access road and the underground networks.

Today, as the town is preparing to re-use more former military zones, the approach is quite different and involves the following stages:

1. reconstitute the history of the place;
2. analyse the evolution of the division into lots;
3. make an inventory of the landscapes and buildings;
4. make an inventory of the sources of degradation and enhancement of the area studied.

On this basis, a comprehensive rehabilitation plan is to be proposed, based, above all, on the constitution of a quality public space.

This notion of public space is fundamental because in conventional planning terms the references are codified (zones, roads with or without services, etc.), whereas in conservation or integrated planning there are no references so long as the historical and typo-morphological analyses have not been completed. The public space therefore cannot be modelled.

Thus Rochefort, a small town of 30,000 inhabitants, is above all an island in the middle of a marshland area, extended by the drainage of its marshland and the multiplication of its defences as far as the mouth of the river, a former oppidum for the defence of the Charente, and a new town organizing the territory via its grid-iron plan. On the strength of this history, the town can reconstruct the masts of its vessels without falling into folklore or addiction to the past.

If the heritage represents elements of shared memory, it can be considered that there are two possibilities: either this memory exists, or it is necessary to reconstitute it (subject of studies) and then share it with the inhabitants. This approach takes precedence over the others, integrated planning then being, at any point of the territory, the way of 'revealing the place' and no longer the simple expression of quantitative development.

Revealing a place is an ongoing process, however. It is a matter of pursuing the historic and heritage analyses and constantly and open-mindedly discussing the knowledge thus acquired. It is also a matter of integrating as well as making possible the necessary modern aspects (cars, utility networks, etc.) in order to avoid internationalizing the place. This means that all the time the problems of re-using buildings, accepting new forms of economic development (tourism in particular) and the various changes, have to be considered from the two points of view of enhancing the memory of the place and of using its capacity to create the future. Then integrated planning gradually becomes a form of expression of the local culture and avoids the three pitfalls of excessive technical specialization, addiction to the past and the loss of individuality of the place.

References

Coussy, B. (1988): 'Le Projet Urbain de Rochefort', *Revue Métropolis,* 82/83 (Septembre).

Coussy, B., Bihel, D. and Gallice, M. (1986): 'Rochefort was not expecting the tourists, but they came all the same', *Town Planning Review*, 212 (March) 71–4.

Coussy, B., Martin-Laval, S., Roze, T., Carrié, B. and Bihel, D. (1998) *Charte Architecturale de Rochefort*, Ministère de l'Équipement, des Transports et du Logement, Direction de l'Architecture et de l'Urbanisme.

ICOMOS (1987): *ICOMOS Charter on the Conservation of Historic Towns and Urban Areas* (Washington Charter).

Other sources of information

Coussy, B. (1987): 'Du rural à l'urbain : 3 lotissements', *Revue Urbanisme*, 219 (May): 120–3.

Coussy, B. and Martin-Laval, S. (1993): 'Faubourgs et Grands Ensembles', *Les Cahiers de l'Habitat*, 20 (Février): 67–71.

Kakha Khimshiashvili

Old Tbilisi, Georgia

Introduction

This chapter serves to provide general information about problems faced in the field of cultural heritage preservation in Georgia, and particularly in the historic centre of Tbilisi, as well as the ways that the Government of Georgia and the Municipality of Tbilisi are trying to solve them.

Georgia, located on the north-east of the Black Sea, was one of the republics of the former USSR. Since 1991, Georgia has enjoyed its independence. Under Soviet government, Georgia was one of the leading republics in the field of conservation due to its rich culture and diversity of heritage monuments. Furthermore, the Georgian people have a special attitude towards their cultural heritage. As one of the most prominent Georgian scholars put it: 'In fact, Georgia has not had a political history during the last two centuries but rather a cultural history' (Beridze, 1995). Consequently, the appreciation of the cultural heritage is essential and important in Georgia.

Conservation was very popular during the 1970s and 1980s. The major body in the field of conservation was the Main Department for Monument Protection. It was responsible for the entire spectrum of a monument's preservation (state policy, priorities, commission, implementation and control) and was entirely financed by the state. This used to be a common situation in the Soviet management system.

The situation has changed dramatically since the collapse of the Soviet Union in 1991. The entire system of the socialist economy, including its management tools, vanished. Owing to the scarcity of resources, the government has been unable to budget for all the works in the area of conservation (as well as for other spheres of life) as it used to in the Soviet period. Georgia is currently in a so-called 'transition period', and is trying to build a new state based on the values of democracy, human rights and moving towards a market economy. This is an ambitious as well as a difficult undertaking. Due to the immense difficulties in this transition period, many spheres of the country's life are in deep stagnation. Whilst Georgia's economy and industry are affected, it is the social and cultural spheres which suffer the most.

The Main Department for Monument Protection still exists but its effectiveness has been drastically reduced because of its budgetary dependence on state subsidies. Consequently it has lacked sufficient financial resources (enough to say that for conservation works for the year 1999 only 0.91 per cent of the adequate budget of 1990 was allocated). In response to this situation the Government of Georgia decided to apply to the World Bank and the Council of Europe (Cultural Heritage Department) for assistance in creating a new system for cultural heritage management.

Thus, the *Cultural Heritage Initiative* (CHI) a joint programme undertaken by the World Bank and the Government of Georgia – was set up by the Georgian President, E. Shevardnadze. The Cultural Heritage Department of the Council of Europe soon joined the programme by developing a Specific Action Plan for Georgia. Through this mechanism the Council of Europe provided technical assistance for the CHI.

The Cultural Heritage Initiative

The CHI programme consists of six components:

1 legislative development;
2 resource mobilization/fund-raising;
3 survey/recording of monuments;
4 community participation;
5 public awareness; and
6 training.

The duration of the programme was set for 18 months, starting in March 1997. Within this programme all components were equally important but the predominant issues have been:

1 Legislative development under which the programme rendered assistance to the relevant bodies for elaboration of a draft of the Law on Protection of Cultural Heritage (Pickard, 2001);
2 Survey/recording of monuments under which the methodology for recording historic buildings was revised and a computerized database created including the development of a 'Cultural Heritage Identity Card' (discussed below) as part of the CHI.

These two components are emphasized: (a) it is impossible to speak about any further development of the process of preservation of cultural heritage without proper a legal basis; and (b) the recording of the immovable heritage is important because it is a keystone for understanding, interpreting and conserving historic buildings. This work is especially significant for the elaboration of integrated conservation strategies for historic centres/

areas of cities and group of buildings. The importance of recording is well articulated in a number of international documents (Council of Europe, 1995; ICOMOS, 1996) In addition, article 2 of the Granada Convention is relevant.

In 1997 it was decided to extend the Cultural Heritage Project, but the focus of the project was shifted towards practical works. For the new project the Government of Georgia requested a Learning and Innovation Loan (LIL) from the World Bank in the amount of US $4,500,000. The contribution of the Georgian Government was US $490,000. At the same time, the Cultural Heritage Department of the Council of Europe has continued its Special Action Plan for Georgia. The CHI was re-registered according to the Georgian law and transformed into the Fund for the Preservation of Cultural Heritage of Georgia (the *FUND*). It is an independent entity or in other words the *FUND* is not included in any of the executive branches of Government of Georgia, neither ministries, nor departments. The *FUND* is responsible directly to the President of Georgia.

This new project consists of two major investment components: (a) an Emergency Rehabilitation Programme; and (b) four pilot sites: (i) Tbilisi – Zemo Kala district; (ii) Uphlistsikhe – archaeological site; (iii) Signagi – a historic town; and (iv) Shatili – an historic mountainous village.

Of the four pilot sites the *Tbilisi Pilot Project* (i) is the most important and significant as it deals with the capital of Georgia and also due to the complexity of problems faced, innovative approaches to be tested there, and the expectations that have been raised.

Historical backgound

A few words about the historic centre of Tbilisi and the Zemo Kala pilot site should be stated. Tbilisi was founded in the fifth century by the King Vakhtang Gorgasali and since then it has been the capital of Georgia. The nuclei of Tbilisi grew up from two fortresses on both banks of Kura (Mtkvari as Georgians call it) river – Narikala fortress, on the right bank and Metekhi fortress, which is located on the high cliff on the left bank of the river. At this point, it is probably the narrowest part of the river (Fig. 6.1). The medieval city consisted of three districts: Kala, Seidabadi and Isani.

One of the most valuable features of the historic centre of Tbilisi is that the majority of its buildings belong to the nineteenth and the beginning of the twentieth centuries (Fig. 6.2). Street patterns reflect the urban structure of the late medieval period. The reason why the architecture is more recent is that Tbilisi was conquered by the Persian Shah Aga-Mahmad Khan in 1795 and the city was totally burnt down. Very soon the citizens rebuilt their buildings using old foundations, and even cellars. This is evidenced not only by historical sources but also by surveys of foundations and cellars that are much older than the buildings themselves. Thus, today, there are many clues to what the historic centre looked like in medieval times.

The Kala district is in the very heart of the historic centre of Tbilisi. Zemo Kala (Upper Kala) together with Kvemo Kala (Lower Kala) used to be the major residential area of medieval Tbilisi. Religious buildings, the

Fig. 6.1 Narikala Fortress on the left bank, and Metekhi Church (thirteenth century) on the site of the former Metekhi fortress on the right. (Photograph: Kakha Khimshiashvili.)

Fig. 6.2 Typical Tbilisi dwelling houses of the late nineteenth/early twentieth centuries (Avlevi Street). (Photograph: Robert Pickard.)

Royal Palace and Royal offices, the Patriarch's and nobles' palaces, 'caravanserais' and other public buildings were concentrated in Zemo Kala. In the centre of the district the only public square of the city was situated. Many buildings, especially public ones, have survived, for example Sioni Cathedral (fifth to thirteenth centuries), Anchiskhati Church (sixth century) (Fig. 6.3) and its belltower, King Rostom's bath and the 'caravanserais'. What is also important is that Zemo Kala, traditionally, has always been a multicultural and multi-ethnic residential area, where Georgians, Armenians, Azeris and Jews lived together, enriching and influencing each other's culture and traditions. Today it remains a multicultural and multifunctional district.

Finally, it should be mentioned that the charm of Old Tbilisi has always been appreciated by the citizens of Tbilisi, guests of the city and especially professionals – architects and art historians. According to a social survey conducted by the *FUND's* sociologists on the re-allocation residences in the Zemo Kala district, the majority of existing residents wish to stay within the historic part of the city. This is despite the poor level of services and facilities as compared to the relatively newer districts where services are comparatively better.

However, many problems remain including a lack of maintenance and groundwaters that are ruining the historic fabric across the whole of the historic centre. These problems are much worse in Zemo Kala. This is why Zemo Kala was chosen as a pilot site within the Cultural Heritage Project.

The policy and planning framework

Current situation in the historic centre

The current situation of the historic centre of Tbilisi is alarming. On one hand, negligence and lack of maintenance has caused intensive destruction

Fig. 6.3 Anchiskhati Church (sixth century). (Photograph: Robert Pickard.)

of historic buildings (many of them have just fallen apart recently – see Fig. 6.4) that changes the urban fabric of the district. On the other hand, equally intensive construction of non-sympathetic new structures is being carried out. There are many reasons for this, the most significant being:

- *Groundwater problems*: one of the major problems faced by the whole of Tbilisi and especially the historic centre is the increasing level of the groundwater table. This phenomenon has caused many structural problems. Surveys show that the source of this rising waters is quite complex, but the main cause is leakage from the water supply and sewage systems, which is washing away soil in many places, thus undermining buildings, roads and underground infrastructure.
- *Ageing and lack of maintenance*: the majority of buildings belong to the nineteenth and beginning of the twentieth centuries. In many cases some elements of buildings have deteriorated. This situation is aggravated by a lack of maintenance, closely linked with the very low incomes of residents.
- *Scarce resources of residents*: average incomes in the historic centre of Tbilisi are extremely low. According to surveys, the average monthly stable income is about US $20 for a family of three. Therefore, the majority of residents are not in a position to maintain their buildings. The situation is aggravated by the fact that in 1992, shortly after the declaration of independence, almost all the total stock of apartments (residential spaces) was privatized, irrespective of their cultural heritage value. As a result, the maintenance of property became the responsibility of owners who, due to their economic situation, were unable to carry out any repair work.

Fig. 6.4 Old Tbilisi, Zemo Kala district – listed building at 22 Shavteli Street that has fallen apart. (Photograph: Kakha Khimshiashvili.)

- *Uncontrollable development*: though the historic centre is in a dilapidated condition it is still attractive for private developers for its beauty and significance. But unfortunately, very often, developers build houses that are not suitable for the historic environment (Fig. 6.5). This should not happen, because developers need project designs approved by the relevant municipal and other bodies. The procedure is strict and theoretically should avoid any significant alterations of townscape and other relevant norms for the historic centre. But in practice, a survey commissioned by the Georgian Parliament and the World Bank, has shown that developers usually 'buy' those permissions and are able to 'acquire' the right to develop (Government of Georgia, 1998). Thus, unfortunately, corruption among civil servants responsible for building permits is high, which actually promotes uncontrollable development in the historic centre of the city.

Tbilisi historic centre and its protection in the Soviet period
The protection of the historic centre takes its roots in the Soviet period. A number of regulatory documents were adopted at this time. But before these regulations were made, some large-scale social and physical alterations of the historic centre took place. Namely, since its sovietization Georgia has undergone huge social changes. This fact is reflected in the architecture and urban characteristics of the historic centre. For example after sovietization,

Fig. 6.5 Old Tbilisi, Zemo Kala district – 'unsuitable' new development. (Photograph: Kakha Khimshiashvili.)

in 1920s and 1930s, the majority of houses were nationalized, apartments were forcibly divided and new inhabitants were settled. This dramatically increased the density of the population and decreased services, as all residents shared the same facilities (kitchen, bathroom). Something similar happened with important public buildings such as the 'caravanserais' located in Zemo Kala district. Caravanserais were designed and built as hostels for merchants accompanying caravans. This was reflected in the special structure of those buildings – the ground floor was used for caravan animals (camels, donkeys); the first floor as a store house for goods; and there were rooms for merchants/guests on second/third floors. In the Soviet period some of caravanserais were converted into residential houses, where one or two rooms were given to a family. All families shared the same facilities.

Furthermore, since private commercial activity was practically forbidden in the Soviet period, all such commercial space was converted into residential space. In practice, this notably changed the appearance of the streets and building façades because, since the late medieval times up to the pre-Soviet period, certain streets were occupied by guilds or specialized merchants (hence some of the present names of streets and squares – Goldsmith Street, Silver Street, Iron Lane, Cotton Lane, Wheat Square, etc.). Craftsmen and merchants lived in the upper storeys, whilst the ground floors were occupied by their workshops and shops, that opened onto the streets with broad openings. In the Soviet period, these buildings were changed into the residential spaces and consequently all the large openings were closed and only the windows were left.

Later, in the 1950s, large-scale reconstruction works were conducted. They were justified by the need to increase traffic flow, which used to be limited in the historic centre owing to the narrowness of the streets. Thus, embankments were built on both banks of the Kura river, new bridges were constructed and some streets were significantly widened (e.g. Leselidze Street). These actions notably changed the urban fabric and the appearance of the historic centre. Some valuable views disappeared, especially as a consequence of the construction of embankments.

But later still an appreciation of the historic centre of Tbilisi, and its values, started to develop. This was achieved through the progress of professional thinking and, of course, by the influence of international documents (charters, declarations, etc.). Thus, in 1978 large-scale reconstruction and restoration works started in Tbilisi aimed at preserving the historic centre (although these works were stopped in the late 1980s and early 1990s at the time of the collapse of the Soviet Union). The works had both negative and positive impacts. Public appreciation of the historic centre and their understanding of its importance and uniqueness increased (especially among residents). However, action was more orientated towards what can be described as 'façadism' (Fig. 6.6), in other words the architects responsible for restoration concentrated on the visible parts of the main streets of the district. Much less was done inside the historic district. The infrastructure and the provision of services were not considerably improved.

Fig. 6.6 Old Tbilisi – 1980s façade restoration. (Photograph: Kakha Khimshiashvili.)

A number of regulative acts and laws were adopted in the Soviet period. They were implemented to manage the historic centre and to set up policy and planning guidelines for city centre preservation. In March 1975 it was decided to declare the historic districts of Tbilisi a 'State Protected Zone'. In April 1977 the law on 'The Protection and Utilization of the Historic and Cultural Monuments' was adopted. This law was relatively progressive for that period – for example, it not only protected historic buildings ('monuments') but also highlighted the importance of the preservation of historic districts as whole units, the network of historic streets, townscape and views of natural beauty. At the same time, the law is a product of the Soviet Regime and thus it did not provide any regulations for controlling work by private developers or other important issues, e.g. tax incentives for owners of historic buildings or investors/grant donations in historic districts.

In 1986–88 the 'list' of protected historic buildings was revised and more buildings were added to the list. There are two categories of listed buildings of significance: *national* and *local*. This 'list' was an important step for the protection of historic buildings. But it also brought a negative side – the existence of the list allowed a loophole whereby unlisted buildings could be built or unsuitable new buildings could be built in the old district. The argument was that historic listed buildings could not be touched, whilst only unlisted, 'insignificant', buildings could be altered or destroyed.

Probably the most important document from the Soviet period was decree No. 76, issued on 29 January 1985 by the Central Committee of the Communist Party of Georgia and Council of Ministers of Georgian SSR 'About Further Improvement of the State Protection of the Historic Part of Tbilisi'. According to this law three protective zones were imposed:

1 Zone of the State Protected Historic Part;
2 Zone of Regulated Development;
3 Zone of Protected Landscape.

In practice these documents can be described as a 'building code' for the historic centre (and it remains in force until the transition to new systems and reforms have been completed). The following paragraphs present this document in more detail:

1 The *Zone of the State Protected Historic Part* is a framework document for the development of the historic core of the centre. The document demands action to preserve the historic fabric of the district, its spatial structure, character, townscape, skyline and landscape. It imposes a special regime according to which:

   - the hydrological situation should be improved and the air basin should be kept non-polluted;
   - buildings that restrain views of the historic part and are bringing dissonance into the historic fabric should be destroyed;
   - buildings that do not have any historic and artistic value may be changed by new buildings that respect the historic fabric;
   - industrial buildings, workshops and storehouses that decrease the aesthetic value of the district, creating cargo traffic and the danger of pollution and fires, should be cleared from the zone.

   The document forbids some construction works such as the building of car parks, garages, engineering infrastructure (power supply poles and wires, transformers, etc.) and street furniture without special permission from the state bodies of monument protection. What is important is that this document also demands that the Municipality should reject applications for permission for civil works (new buildings, major repair works, landscaping, etc.) that are not agreed upon by the state bodies of monument protection.

2 The *Zone of Regulated Development* can be described as a surrounding area of the historic centre. In this zone the construction of new industrial buildings, storehouses and other structures which create a danger of pollution, fire and extra cargo traffic is forbidden. Also forbidden is the building of new motorways, bridges and other engineering structures that may change the image of the historic centre. In cases of an over-riding necessity for such works, all project designs should be agreed with the state bodies of monument protection. The document demands the preservation of the historical urban fabric and its fragments, and the improvement of the visual qualities of the historic part of the city. For this, a number of places from which panoramic views of the historical centre can be seen, should be defined and preserved. The document protects the 'background urban fabric' from the development of new structures that can damage visual qualities. Finally, as in the case of the *Zone of the State Protected Historic Part*

the document also demands that the Municipality should reject applications for permission for any civil works that are not agreed upon by the state bodies of monument protection.

3 The *Zone of Protected Landscape* protects the natural environment of Tbilisi's historic centre and its surroundings. According to the document, natural landscape (including mountain slopes, forests and green plantations) should be protected. It is forbidden to build structures within this zone. If there is a need to build a structure it is necessary that the project design is agreed, again, with the state bodies of monument protection, as in the case of other zones.

The Tbilisi pilot project

Whilst the majority of these regulatory acts remain in force, including the decree of 1985, unfortunately they are not, in fact, operative, and at the moment are not used in their whole capacity. The Georgian Government decided to include the Zemo Kala district, Tbilisi, as a pilot site in the Cultural Heritage Project in order to improve the situation in the historic centre and to preserve its qualities, as well as to test new strategies and tools for the rehabilitation of the district. If the pilot project is successful it can be used as a good exemplar for other historic districts and even whole historic towns.

It was decided to start planing actions for the rehabilitation of the Zemo Kala district from its inventory. Thus a methodology for recording the immovable heritage was developed, including a specially designed questionnaire, which is universal for the urban and rural environments and separately standing monuments. In this context it should be noted that the Council of Europe's experts provided significant assistance.

Inventory of the historic centre of Tbilisi

A significant number of buildings were recorded in this district during the Soviet period. But it was only the listed historic buildings that were recorded. The *FUND's* experts used a different approach by undertaking a so-called 'blanket coverage', in other words *all* buildings were recorded regardless of their historic and artistic values. The reason why this approach was used is that experts have indicated many times that the main value of Tbilisi's historic centre is its urban fabric, townscape and streets network derived from medieval times. Apart from the general need to gain this information for management and planning reasons, it was also deemed necessary to have information about the whole district. This has helped to develop a clear picture about the physical conditions, ownership and values expressed in the district. All this information has been recorded on a database through means of an 'identity card' and will be used to assist the process of developing 'integrated conservation' management tools.

Cultural heritage identity card

The new inventory database provides a system of recording information on buildings, groups of buildings, and sites according to 25 descriptors. These may be described as follows:

- *Name and cross-references* – includes the name of the building, group of buildings, a site code, identity number and compilation data and cross-references to other buildings, monuments, relationships and to identity concerning archaeological, environmental ethnographical and other information including graphic and photographic records, textual sources and bibliographic references.
- *Visual material* – includes graphic and photographic records.
- *State of preservation* – includes preservation type, grade and date of listing.
- *Location* – includes administrative definitions (State, Region and *Sakrebulo* or local administration), historical and geographical definitions (historical province – Kakheti, Imereti, Ratcha; historical sub-region or 'microstate' – Kiziki, Kudaro, Samurzakano; and non-administrative geographic units – Tedzami gorge, Mukhrani valley), address details, cartographic cross-reference and cadastral reference.
- *Owner/property* – includes the owner of the building (according to controlled vocabulary: state organisation/name/church, public organization/name/private person or group of persons, mixed ownership) and relevant dates of ownership.
- *Function* – includes present and former functional category (religious, civil, industrial, etc.) and type (church, synagogue; theatre, circus; factory, etc.)
- *Date* – information is categorized by period/era, century, chronological periods.
- *Persons and organizations associated with the site* – includes the name of the person or organization and role in the site history (architecture, engineer, mason, customer) by date.
- *Material and construction* – information is recorded on walls and roofs.
- *Historical and architectural résumé* – a compendium of relevant information
- *State* – includes information on the existing situation through a general evaluation (positive, negative), general characteristics (destroyed, damaged), brief description, history of damage and restoration, reasons for damage, natural situation (vegetation, stone erosion, subterranean waters, influence of climate, earthquake, etc.), human interference (whitewashing, new inscriptions, non-correct repair, emergency situation of water sewerage system, influence of neighbouring buildings, nature pollution, fire, etc.) and risks associated with the site.
- *Environment of the site* – includes details on the architectural environment (connection of the site with the environmental architecture – buildings, yards, streets; place in architectural complex) and the natural environment.
- *Description of the site* – is carried out according to special separate schemes for buildings, for groups of buildings and landscape.
- *Historical data* – relevant historical information.
- *Inscriptions* – separate points, marking the exact location (lapidary inscriptions, colour inscriptions and graffito).

- *Information about the pieces of art connected permanently with the site* – including relief, frescoes, etc. – all items are described by place, name, author, quantity, date of construction and by a brief description (only if the site possesses historic and artistic values).
- *Information about small architectural and artistic forms* – for free standing monuments, ventilation constructions, poster columns, etc. – including a description of the place where it is situated, name, quantity, author (house, factory, etc.), the date of construction, and a brief description (only if the site possesses historic and artistic values).
- *Information on non-architectural equipment and machinery connected with the building* – for stationary mounted furniture and different industrial equipment, organ, etc. – each item will be described (similar sites – in one group) by place where it is situated, name, quantity, author (house, factory, etc.), date of installation, and a brief description (only if the site possesses historic and artistic values).
- *Information on movable items kept in the building* – and which possess historic and artistic values (furniture, vassal, carpet, pictures, icons). The basis for recording such items is that defined by the Getty Information Institute (Thornes and Bold, 1998).
- *Historic and artistic value* – relevant information on the identified value.
- *Information on technical standing of the building* – including existence of power supply, natural gas, water supply, sewage facilities and lightening protection.
- *Demographic and social information* – including a brief historic and demographic excursion (population details, ethnic constituency, (un)employment information and information on pensioners).
- *Physical and geographical conditions* – including climatic conditions, geology and relief, and information of soil characteristics, flora and fauna, landscape description, etc.
- *Infrastructure* – including details on roads, urban transport, airports, post offices, police services, accommodation facilities and communication systems in the vicinity of the site.
- *Additional information* – any other relevant information.

Geographical Information Systems (GIS)

The *FUND* purchased a digital map of the district. It has pioneered the use of digital maps not only for cultural heritage inventories, but also, in general terms, for the benefit of different municipal departments that are now beginning to use them. A digital map of the UTM Projection utilizing the WGS-84 Co-ordinate System with 15 cm adjustment was used as the basis for developing thematic maps, using GIS tools (MapInfo Software). Alphanumeric and object-oriented layers were added, so that 'thematic maps' can be prepared.

Thematic maps reflect a number of features and give a vivid picture of the current situation of the building stock of the district. The historic value, physical condition, date or period of building, ownership, façade material, roofing material, existence of balconies, material of balconies, statues of protection, etc., are among the features recorded.

In addition, GIS gives an opportunity to conduct enquiries combining a number of features so that buildings of specific interest can be examined. For example, if one is looking for a building of average artistic value dating from the nineteenth century, in good physical condition, with a wooden balcony, the system immediately provides a visual answer as to whether there was such building or not.

A big advantage of using GIS is its open character; in other words it is possible to add new information (layers) to the already existing layers. Information can be of different character such as density per building, infrastructure services (water and gas supply, toilet facilities), network of water supply and sewerage systems, residents and tenants income, geology of the district, etc.

Thus the database inventory and the thematic maps are proving to be excellent tools for analysing and forecasting development in the district. Through these aids a solid foundation of information can be accessed for developing integrated conservation plans. The final product should be a plan (strategy) for the sustainable rehabilitation of the Zemo Kala district.

All this information is available to the public and municipal departments. The municipal departments have already expressed a strong interest in the maps.

Management and regeneration action

Preliminary phases

Having completed the inventory of the Zemo Kala district it was decided to test new integrated conservation approaches and start rehabilitation works. The inventory in conjunction with the GIS facility helped to develop a strategy for civil works. Within the budgetary limits of the *Tbilisi Pilot Project* it was decided to divide civil works into two phases.

At the same time a number of surveys and studies were conducted either by the *FUND*'s relevant units or by the request of the *FUND*. First of all it should be mentioned that the *FUND*, together with the Georgian Government, requested assistance from the World Bank in assessing the groundwater problem and preparing a feasibility study for water management in the historic centre of Tbilisi. This study was financed by the Dutch Trust Fund, and was conducted by a Dutch company together with their Georgian counterparts. Having the results of this study, which gives a clear picture of the extent of the problem and a cost estimate for the water supply and sewage system rehabilitation works, it is easier to reach potential investors.

The social survey of the district conducted by the group of sociologists on behalf of the *FUND* provides interesting material for analysing the current situation. In this respect, residents' attitudes towards the cultural heritage are very important. The survey showed that appreciation of the cultural heritage and understanding of the importance of its preservation is quite high among residents of the district. Another important result of the survey is in its social impact. It is interesting that residents highlight the fact of the survey itself. As they explained, it is the first time that official interest has been taken of their opinion. They compare this event with their Soviet experience when nobody was interested in the opinion of local residents, regardless of the type of problem – preservation of cultural heritage, infrastructure improvement or any other works. Thus, this fact has had its own impact and created a certain credibility towards the *FUND* and its activities.

Under phase one of the project more conventional measures of rehabilitation of the historic centre have been planned. Namely, it was decided to concentrate resources on important public buildings and spaces on one hand and to improve the living conditions of local residents on the other. Hence two public gardens, the creation of landscape improvements, and rehabilitation works for three museums (Baratashvili Museum, Tbilisi Ethnographic Museum and State Art Museum) were chosen. As for works in residential areas – the Emergency Repair Works Programme (or as it is often called: the 'Neighbourhood Fund Programme') was launched.

The aim of this programme was threefold:

1 to improve the living conditions of local residents;
2 to improve the physical condition of historic buildings; and
3 to gain credibility from the community towards the project and the actions of the *FUND* in order to create the favourable atmosphere for Phase two works.

Under the Emergency Repair Works Programme, owners of listed properties of historic significance can apply to the *FUND* for grants to carry out urgent repairs to buildings. The amount of the grant is not large: the maximum amount offered per unit is US $1,500. In the case of a number of owners/families living in the same building presenting one collective application for repair works in relation to the entire building, then US $4,500 can be granted (the amount of three individual grants).

It is important to note that since the Emergency Repair Works Programme is aimed at the assistance of the vulnerable part of the society, the applications of these groups are given priority. Collective applications are given priority as well – to stimulate the community thinking of the society.

Phase two of the Tbilisi pilot project is the most important component. Under this component it is envisaged that a sustainable process of rehabilitation of the Zemo Kala district will be launched. To commence phase two, the *Tbilisi Pilot Project* 'Framework Document' was elaborated and adopted by the Fund, the World Bank and the Council of Europe (Fund for the Preservation of Cultural Heritage of Georgia, 1999). The document was presented to, and accepted by, the Mayor of the city. Because of the importance of the document, which identifies the main concept of the district rehabilitation process, further details are provided in this chapter.

The second phase of the project is also closely connected with previous works of the *FUND*. Recording the district and conducting surveys and studies are especially important. A concrete action plan is to be developed at the beginning of 2000 and will contain many new innovations for Georgia, such as participatory planning tools and principles of integrated conservation.

An important part of the future work is a revision of existing building codes and other regulatory norms dealing with the preservation of the historic centre of Tbilisi. Recommendations will be prepared and submitted to the municipality and other relevant bodies for their adoption. It is hoped that if the recommendations are fully adopted, the level of 'uncontrolled' development will significantly decrease. Also some institutional adjustments will be made which will improve the performance of municipal departments involved in the preservation of the historic centre of Tbilisi.

It should be emphasized that the project is supported not only by the international organizations directly involved in project realization (the World Bank and the Council of Europe) but local organizations as well, such as ICOMOS GEORGIA and the Union of Urbanists of Georgia.

Conceptual framework of the Tbilisi pilot project

Preparation stage

The 'Framework Document' identified that the regeneration project should be developed according to two stages – in order to accomplish timely and cost effective demonstration results. It was determined that sites to be designated for action should be identified on the basis of social, economic and physical surveys of Zemo Kala, followed by a careful assessment of the conditions of infrastructure networks and urban services provision, and an analysis of area development opportunities. A systematic survey and assessment of buildings' structural and maintenance conditions will be subsequently undertaken. A set of feasibility studies will be carried out in the first stage. These will provide reliable data and indicators about the local population, economic and social activities and cultural opportunities in the area. This basic information was identified as being necessary for the selection of pilot sites and the preparation of the preliminary programme design. The document identified that a participatory planning approach should be initiated at the outset with the residents of the selected areas for development of pilot projects. For development of architectural preservation and urban design concepts, a continuous process of public consultations will be carried out at the neighbourhood level and, at the same time, with key city stakeholders.

Community involvement in urban design and preservation decisions can generate long-term benefits, by resolving differences in the planning process, step by step, before the work is consolidated into proposals. Moreover, this approach has been deemed essential in order to build strategic partnerships between the public and private sectors, and hence to mobilize local resources for project development and its sustainability.

Scope of work
The two stages were identified as follows:

STAGE 1

1 Data gathering, compilation and analysis of documents, related to urban legislation (land-use planning and zoning), building code and municipal regulations concerning urban services, and complementary research of archival documentation on listed historic buildings to be restored.
2 Engineering and urban services survey in the intervention area, including environmental sanitation, conditions of infrastructure networks and provision of urban services, pedestrian and vehicular circulation (parking spaces and service areas).
3 Develop a participatory planning approach with the residents of Zemo Kala, stakeholders, experts and city officials, by organizing public consultations and neighbourhood meetings.

STAGE 2

4 Prepare a development framework plan based on the findings of sector surveys and studies (see Stage 1). The report should contain recommendations for action. This development plan should examine the ongoing public and private investment activities and review the regulatory and normative plans, i.e. the existing master or development plans of the various municipal departments for that district. This plan must also indicate the potentials and opportunities for sustainable protection management of Zemo Kala. The key chapter of this plan should be the guidelines for development in the historic district that will be submitted to and approved by the City Council – see point 5.
5 Prepare urban design guidelines for Zemo Kala. The guidelines may include parameters for new construction and standards for the renovation of buildings, courtyards and alleys, and establish the relationship between built areas and open space. It may also determine the adequate land use plan for the area, i.e. the maximum allowed construction volume on any given parcel of land (according to floor area ratio and envelope of the building), and density in different sectors. In addition, the guidelines may recommend regulations for commercial and residential areas, and improve circulation in the neighbourhood by separating pedestrian and vehicular flows and designating parking areas.
6 At the same time, based on the technical findings and surveys conducted, pilot project intervention sites will be selected.
7 Prepare detailed terms-of-references (TOR) and tender packages for the identified engineering, landscape and architectural projects. The TOR should contain a detailed description of the works to be performed, plans, details, implementation schedules and timeframe.
8 Prepare and test a participatory monitoring and evaluation system.
9 Implementation of engineering and restoration works.
10 Ex-post evaluation and assessment of the project outcomes and results.

Implementation strategy

The proposed implementation strategy for the Zemo Kala pilot project hinges upon two working principles. First, at the institutional level, direct involvement and active co-operation of key municipal services departments in the realization of specific project activities. Second, at the community level, securing residents' and private sector participation in the preservation and revitalization of the project designated area. In short, the strategy aim is to enable residents and stakeholders to join in a collaborative effort, and partake in the protection and conservation of historic buildings, hence increase overall resource inputs while improving the conservation management approach of the historic district.

At the institutional level, it was identified that a 'Co-ordinating Council' should be established and meet on a regular basis to deliberate and facilitate project inter-sectoral co-ordination and implementation. The *FUND*'s role is to convene and organize council meetings, prepare the agenda and issue and circulate the proceedings of meetings held. Council members represent key municipal departments (among others, chief architect, historic monuments protection inspectorate, public works and utility services), representatives of the private sector and community leaders. The members of the Co-ordinating Council will then deliberate on technical and implementation aspects of the project and also those pertaining to the legal and regulatory frameworks.

At the local level, a series of public consultations and meetings will be held prior to drawing up the preservation plans. Securing support from residents and stakeholders, i.e. the project beneficiaries, was considered to be fundamental to the success of the initiative – residents must 'buy in' and 'own solutions' from the outset.

In order to meet the project objectives in a timely fashion, a work plan covering the first year of activities is to be prepared. This plan should be presented, reviewed and validated by the project Co-ordinating Council and in consultation with residents, as to its implementation timeframe, itemized resource allocation, and expected inputs from city agencies and private sector participation.

Conclusions

Cultural heritage is an important factor in sustaining national unity and in fostering social and economic development. It may also contribute to the ongoing efforts to complete Georgia's transition to a market-led economy. The built heritage is an invaluable asset that, if properly maintained and preserved, can generate social and economic benefits to local communities – generating employment and rent, and thus improving living conditions. It can be the means by which to engage residents in economic activities in keeping with their traditions and culture.

The objective of the proposed pilot project is the preservation of Zemo Kala's historic architecture and cultural and religious heritage, expressed in a variety of buildings. Preservation also means improving the living conditions of its residents, by maintaining the neighbourhood identity, community cohesion and the unique architectural ensemble of residential buildings and open space. This has been the main reason for launching the rehabilitation of Zemo Kala – through a conservation and revitalization effort. It is noteworthy that the basic assumption of this project is to test a new preservation approach based on promoting and brokering partnerships among public and private sectors and the community. This approach calls for active participation by the local residents and stakeholders in all stages of the project development process. The pilot project should be carefully assessed in terms of the quality of its results, i.e. its cost effectiveness, quality of works performed and beneficiary satisfaction.

As a result of project implementation, the living standards of the local residents and other stakeholders of Zemo Kala are expected to improve. The revitalization of the neighbourhood and preservation of its architecture is a complex process that requires the adoption of a balanced strategy. An important outcome of the pilot project is the enabling of new economic activities and expanding employment opportunities in the area. This may also increase property values in the district and consequently raise private resources for the preservation of the historic ensemble of buildings and their maintenance.

In the long term, a new dynamic process may take place that will encourage new activities including tourism. Through investment in historic properties and the establishment of economic activities the flow of visitors to the district is expected to increase. It is hoped that tourists can be encouraged to experience the cultural heritage of the capital of Georgia (Fig. 6.7).

Both the revitalization and preservation process should be based on a sustainable development approach. Thus, the pilot project may demonstrate

Fig. 6.7 The Hotel Tbilisi which was destroyed by fire in post-independence conflicts in December 1991 and is now undergoing rehabilitation to cater for future tourist activity. (Photograph: Robert Pickard.)

a new and viable approach to the preservation of the neighbourhood, which could be replicated in other parts of the historic district. Moreover, lessons learned from the project implementation process may demonstrate ways and means for the adoption of a sustainable preservation strategy for the whole of old Tbilisi and other historic towns of Georgia as well. The experience of developing institutional co-ordination mechanisms, as established under this project, will be an important asset in the development of such strategies.

As was mentioned at the beginning of this chapter, Georgia is in a 'transition' period, between the former Soviet regime and the adoption of the values of democracy, human rights and the development of a market economy. Indeed this country is at a crossroads in its social and economic life. It is hoped that the *Tbilisi Pilot Project* will be a little step towards the development of a civil society in Georgia and towards joining the European family of nations.

References

Beridze, V. (1995): Speech on the session of the Institute of History and Art of Georgia, Academy of Science.

Council of Europe (1995): *Core data index to historic buildings and monuments of the architectural heritage*, and, *Recommendation R (95) 3 of the Committee of Ministries of the Council of Europe to member States on co-ordinating documentation methods and systems related to historic buildings and monuments of the architectural heritage.*

Fund for the Preservation of Cultural Heritage of Georgia (1999): *Tbilisi Pilot Project Framework Document.*

Government of Georgia (1998): *Briefing Notes. Coordinating Reforms in the Public Sector: Improving Performance and Combating Corruption.* A workshop organized by the Government of Georgia with the assistance of the World Bank, 21–23 June, Tbilisi, Georgia.

ICOMOS (1996): *Principles for the Recording of Monuments, Groups of buildings and Sites* – Text ratified by the 11th ICOMOS General Assembly, held in Sofia, Bulgaria, 5–9 October 1996.

Pickard, R. D. (ed.) (2001): *Policy and Law in Heritage Conservation*, Conservation of the European Built Heritage Series, Spon Press, London (see Chapter 6: Georgia by Simonishvili, M).

Thornes, R. and Bold, J. (eds.) (1998): *Documenting the Cultural Heritage*, Getty Information Institute.

Silvia Brüggemann and Christoph Schwarzkopf

Erfurt, Germany

Introduction

When thousands took to the streets in 1989 in Thüringia to protest against the sorry state of affairs in the German Democratic Republic of East Germany (GDR), one of their major concerns was to preserve the historical buildings. These were not only generally at risk in the city following decades of neglect, but, at the time, were in imminent danger of being demolished due to the proposal to pull down the city's old quarter, the Andreasviertel (Andreas Quarter) of Erfurt. Tens of thousands formed a 'citizens' bulwark' along the former inner city wall to oppose the historic old quarter's decay and its threatened demolition symbolically.

The history of the city of Erfurt goes back more than 1,250 years. Today's old quarter developed from about the year 1000 out of the settlement at the ford over the River Gera where the *via regia*, which led from the Rhine to Russia, crossed other important trade routes. Initially, the Gera skirted the settlement in a semi-circle, whilst *Domhügel* (Cathedral Hill) and *Petersberg* (Peter's Mount) formed its north-western boundary. In the twelfth century, a first, 8km-long stone wall was built. The city had its heyday in the fourteenth century when it was able to found a university of its own. The city's best known historical monuments – the *Dom* (Cathedral) and the *Severikirche* (Severi Church) – date from this period, as do other churches in the old quarter. New fortifications were built in the fifteenth century around the city that had expanded. No secular building has survived intact from this period, however, except for the stone structure *Krämerbrücke* (Krämer Bridge) which was built in 1325. Otherwise, the oldest, more or less intact structures, date from the Renaissance.

Even if this was no longer the heyday of the city, some of these structures are nevertheless characteristic of the aspect of the old quarter. Whereas at its height the city was treated as a free, imperial city, it had to submit completely to its sovereign lord, the Archbishop of Mainz, in 1664. A vicariate was set up in the city. Following the *Reichsdeputationshauptschluß* (decision of the Deputation of the German Estates compensating German sovereign princes for losses of territories ceded to France) of 1803, the city became Prussian and remained so until 1944/45, except for an interval early on during Napoleonic rule.

It was only after the fortress of Erfurt was disbanded in 1870 that the city was able to expand. Within the ring formed by the former city wall, enclosing an area of some 150 hectares, important individual buildings have been preserved, in particular Gothic churches and secular buildings dating from the Renaissance and Baroque periods. The structure of the city's ground plan and a significant number of older blocks of buildings in this area have also been preserved.

This observation must be qualified, however: on the one hand, numerous older buildings were demolished in the last century to make way for commercial buildings which were modern in their day – and which today are objects of conservation. On the other hand, urban renewal during the first decades of the twentieth century pursued a totally different course from today. This was based on a building line plan, through which whole blocks of buildings in the old quarter were to be pulled down and replaced by new ones, so as to improve the disastrous hygiene situation in the city's narrow streets. These ideas which, after World War II, literally 'forged ahead' in both East and West were consequently not new to the GDR either. It was merely their implementation that was simplified here because no owner had any means of protecting his property from the covetousness of 'modern' or 'socialist' urban planning.

At the same time, official conservation authorities existed even before 1989. In those days, however, they confined themselves – where they saw the city ground plan as worthy of preservation – to preserving individual structures. Given a centrally managed economic policy, any more than this was impossible, for the latter envisaged 'refurbishing' the old quarter with concrete slab buildings. This could only result in the abandonment of the city ground plan, as indeed was the case with the new building in the northern part of the old quarter up to 1989.

These changes went hand in hand with a traffic concept that envisaged sacrificing the *Andreasviertel*, a part of the old quarter that had hitherto been spared alterations of note for some 200 years, to a broad highway. The deliberate neglect of the historic buildings aided and abetted this scheme.

The planning and policy framework

Emergency salvage work since 1990 – Thüringian initiative to salvage buildings

The circumstances described gave rise to the situation prevailing in 1989/90 when the climate for preserving the old quarter after the demise of the GDR was favourable in principle. In this situation, the salvage work made a good impression. 'First aid' came from the Rhineland-Palatinate at the beginning of 1990. This enabled a first group of buildings to be repaired; others were to follow.

In October 1991 *Land* (autonomous state) Thüringen launched the 'Thüringian Initiative to Salvage Buildings' to halt the decay in individual buildings. Aid was to be granted on condition that the cities – Erfurt too – undertake to issue redevelopment statutes under the Building Code, following appropriate local authority or town council resolutions. Even if this was on occasion belittled as a 'Roofs on Ruins' campaign, this Initiative to Salvage Buildings did serve as a starting point for active urban redevelopment. In this way, some 300 endangered buildings were examined over a two- to three-year period, and their roofs and loadbearing structures repaired, halting their decay. This was important in view of the fact that, at this period, ownership questions had to be clarified. So-called emergency salvage work could also be carried out on housing where the question of ownership was still open.

Town and country planning

The town plan as *Bauleitplan* (Development Plan), which covers the entire territory under a local authority's jurisdiction, is best able to provide information about its planning objectives, where treatment of the man-made landscape, and inclusion of the old quarter in the development of the city as a whole are concerned. The remarks on the preliminary draft of 1996 define these objectives. They state: '2. The city of Erfurt is embedded in a landscape whose character must be preserved. The slopes characteristic of the city must remain recognizable, and the valleys converging on the city must not be obscured by buildings. 3. The Erfurt townscape is characterized by the aspect of the more than 1200 year-old city and its surrounding fringe of villages and suburbs. The building and spatial structure of the villages and suburbs must be preserved in the city's future urban development too' (FNP, 1996).

In the 'Basics of the Explanatory Report on the draft FNP (Town Plan)' of 1999, this is explained. It states: 'The villages must not be allowed to form an unbroken line of buildings with the city' (FNP, 1999).

In accordance with the requirements of the *Baugesetzbuch (BauGB)* (Building Code), the Town Plan indicates historical monuments for information purposes. However, they are only plotted on the map of the old quarter (Wittich, 1999). Commenting on the importance of the monuments, the preliminary draft (FNP, 1996) remarks as follows: '4. Erfurt is amongst the oldest cities in Germany and has one of the largest medieval inner cities in which there are a large number of historical monuments. The centre of Erfurt exudes great appeal and radiates a lively urbanity. It must retain its appeal even as economic structures change.' The peripheral areas are to be linked with the city core by ribbon-like urban development.

The old quarter is not shown as a core area but as an '*allgemeines Wohngebiet*' (general residential area) *(Baunutzungsverordnung sections 1–4, 7)*. This is in effect to profess that the mixture of residential accommodation and business, local government and commerce is intended politically. It is not without symbolism that the Premier of Thüringia resides in the old quarter.

Binding planning for the old quarter

In 1991 five sections of the old quarter were designated as redevelopment areas. In mid-1992, following preparatory studies on the extended old quarter, a redevelopment statute for the whole of the old quarter covering an area of some 220 hectares was adopted. In 1993 this was extended to include the area at the *Karthäuserstraße*, south of the old quarter.

In addition to refurbishing and modernizing the buildings and infrastucture located here, a major objective of redevelopment is to create urban harmony between the historic old quarter, the buildings on the periphery of the city dating from the *Gründerzeit* (period of rapid industrial expansion 1871–73) and, subsequently, the new housing estates dating from the 1960s, 1970s and 1980s. By 1994, a framework plan had been drawn up by the municipal authorities as a basis for redeveloping the old quarter.

At present, there are no binding plans for the old quarter in its entirety. Precise guidelines for basing individual decisions, which the city's building authority is constantly having to make, can be found in a number of documents that have been available since 1992. These include the '*Erhaltungssatzung für die Altstadt von Erfurt*' (Conservation Statute for the Old Quarter of Erfurt) under Section 172 *BauGB* (Building Code), the *Amtsblatt* (Official Gazette) of 24 June 1992, and the '*Ortsgestaltungssatzung für die Altstadt von Erfurt*' (Local Design Statute for the Old Quarter of Erfurt) (Amtsblatt 25 November 1992) under Section 83(1), no. 1, 2; (3), p.1. *ThüBauO* (Thüringian Building Code).

Design and conservation statutes

Section 2 of the Design Statute lays down that construction work must not impair 'the urban design of the area and its importance in conservation terms.' This is the object of the following directives:

- the line of the façades facing the street, as well as any departures from it, must be retained;
- the street façades of new buildings that exceed the plot boundaries that applied in 1945 must be designed to accord with the frontage widths that existed then.

In anticipation of the *following* directives, Section 2 provides – out of a justified concern that architectural uniformity would otherwise be prescribed – that exceptions are possible.

Under Section 3, buildings with a pitch of roof of 15–65° are to be built with the eaves running parallel to the street, the pitch of roof having to be 'adjusted to match its surroundings'. Red tiles are prescribed as roofing, slate and metal only being permitted for repairs to existing properties. The size of roof super-structures is also regulated, and individual aerials and satellite dishes are prohibited where visible.

Section 4 gives detailed instructions on the design of façades. These are to be divided up into a ground floor, an upper storey zone and a garret zone. Neighbouring buildings or building sections, as required by Section 2, must preserve the character of a single house in the façade by varying the design elements. Unbroken expanses of façade must account for 50–80 per cent, whilst the openings must be vertical in format. Even if the ground floor façade is entirely open in design, the loadbearing structure must be visible in the façade. The walls must be given a plain coating of plaster. A fresh coat of paint must be applied to accord with restoration studies and their findings. Further provisions relate to doors and windows (Section 5) and advertising facilities (Section 6).

Erfurt does not have an actual design manual such as other cities use and these very restrictive provisions make it almost imperative to issue separate Local Plans as binding development plans for sizeable building projects in the old quarter. This means that 'departures' from the Design Statute do not have to be approved as exceptions. The Local Plans in respect of the underground garaging on the Petersberg, a considerable challenge in conservation terms, and the Federal Labour Court on the Petersberg, assumed particular importance. Erfurt does not have an explicit Monument Conservation Plan as would be possible under the Thüringian Law on the Protection of Historical Monuments. The matters to be regulated are mentioned in the Design Statute in general terms.

Further questions are a matter for the permit application procedure under the Thüringian Law on the Protection of Historical Monuments. Section 2 of the Conservation Statute provides as follows: 'To maintain the structural character of the old quarter of Erfurt resulting from its structural design, a permit must be obtained before buildings in the area to which this Statute applies can be demolished, altered or put to a different use, or before new structures can be built.' The issuance of this Statute was a precondition for Erfurt's inclusion in the Federal Programme entitled '*Städtebaulicher Denkmalschutz*' (Preservation of Historical Urban Monuments).

Analysis of the old quarter

The object of the salvage work, for which there was no overall plan at first, was to stop the building fabric from decaying further so as to ensure that there would still be historical buildings for inclusion in planning at a later stage.

The *Stadtplanungsamt* (Municipal Planning Office) drew up a Framework Plan for the old quarter by 1994. This relates to the entire old quarter of 200 hectares, shown since 1992 as a redevelopment area under the so-called *Vereinfachtes Verfahren* (simplified procedure) pursuant to Section 142 *BauGB* (Building Code). The municipal authorities have less influence here than under the so-called 'Vollfahren' (full procedure). Detailed inventory of the condition of the building fabric and its development potential has been made in this connection only for the areas under full procedure. Such an inventory is normally drawn up as part of the 'Preparatory Studies' under Section 141 *BauGB* (Building Code). The object of these studies is not to draw up a detailed inventory of historical monuments, however, but to ascertain urban problem areas where design, social and economic questions are concerned, as well as questions of building substance. The studies form the basis for deciding whether an area is even to be designated a redevelopment area.

This has not been an issue either in Erfurt or in the other cities of Thüringia or East Germany. That is why the preparatory studies had more to do with social questions than with the buildings. Scrutiny of the results of preparatory studies conducted elsewhere casts doubt on the validity of the studies in conservation terms anyway, as the planning bureaux conducting them do not necessarily have the requisite skills, and, on the other hand, the criteria for data to be collected have not been defined. The results of such studies are only seldom analysed as in Königsberg in Bavaria. For example, a well-known Munich architect conducted preparatory studies there, and the results were found to be extremely questionable when examined as part of a more recent research project. This project not only questioned the point of dividing buildings up into ages spanning 100-year periods; it also found after conducting scientific tests on the age of buildings (dendro-chronological, structural and archival) that the ages given for the buildings in the preparatory studies were in many cases incorrect (Imhof, 1993). This is not surprising as these were only conducted visually. It can therefore be appropriate to dispense with a superficial analysis that pretends to be more, in order not to create the fiction of conservation validity, and to carry out more precise studies only in specific cases. It is also regrettable from a scientific point of view that no systematic building research is conducted (see, *inter alia*, Reinhardt, 1999).

Since 1997, data on the cellars in the old quarter is admittedly being compiled systematically, and this in turn permits conclusions to be drawn on how conservation should proceed in the entire area. This research is a continuation of work on a 'Cellar Register' begun in the 1980s (Altwasser, 1999: 19). The fact that the work is being carried out as a job creation scheme funded by the Bundesanstalt für Arbeit (Federal Labour Exchange) highlights the chief problem. This is that it is politically almost impossible to put across that large-scale research is being conducted and that promotion funds cannot therefore be used for the building fabric itself. In this connection, it is almost immaterial in what proportion the funds are allocated.

The situation with regard to the refurbishment of the infrastructure of Weimar (also situated in the *Land* Thüringen, near to Erfurt), in which an enormous amount of aid was invested, in preparation for it being the European Cultural City for 1999, was a little different. Here, architectural studies were commissioned block by block in order to have an appropriate planning basis in conservation terms. One of the results of this investigation was that financial savings, many times the cost of these studies, could be made by giving an avenue dating from the *Gründerzeit* a bituminized surface. By adding chalkstone chippings the surface is made a little lighter. The research had found (Brüggemann and Schwarzkopf, 1997) that this was the closest match to the original macadam surface, rather than paving it as was originally planned.

The Framework Plan for the old quarter
The Framework Plan for the old quarter of Erfurt, completed in 1994, should be seen against this backdrop.

One of the objectives of urban planning since its inception this century is to have the city crown, composed of Petersberg, Dom and Severikirche, continue to serve as a highlight. In so saying, the Framework Plan accepts the buildings of the 1960s that mar this silhouette. These include the 'socialist city wall' (apartment blocks of 11 floors) on the southern edge of the old quarter, which has a negative effect on the city's micro-climate, and also today's Radisson Hotel and other high-rise buildings to the east of the old quarter.

Remarking on the adaptation of the old quarter to present-day needs, an official report states: 'Like every city, Erfurt is subject to continuing change. New functions (modern forms of shopping, services and recreation) are not to disrupt the existing structure. That is why the size of plots and scale of buildings are to be maintained as a matter of principle and, where necessary, gaps filled with new, modern buildings' (see Homepage, listed in the References at the end of the chapter). To be sure, major inroads into the city's existing structure have already been permitted. Moreover, it is questionable whether it is even possible to preserve the traditional plot structure without paying the price of commerce moving out into the countryside.

A further Framework Plan objective is to preserve the courtyards still in existence and turn them into public spaces, at the same time creating new ones. This was actually an urban planning goal in the days of the GDR when the courtyard of the building at Marktstraße 21 was opened up to the public. One might also ask whether integration of courtyards is worthwhile with a view to preserving historical spaces, whilst basically creating a false picture of history: a tidy courtyard turned into a beer garden never existed in Erfurt in this form. The courtyards housed workshops, storage space, and offices. It is nevertheless a fact that conversion of the courtyards into semi-public open spaces improves the quality of urban life.

'Public spaces', i.e. streets and squares, are to be 'systematically developed for public use by the townsfolk'. An end must be put to 'the dereliction of urban land' (see Homepage), car parking is largely to disappear from the townscape. Unfortunately, most of the really historic urban ground was removed at the time of the GDR, or earlier. Today, really old paving can only be found in a few places in the town centre. Usually new paving is laid that is not historical in design, as was recently the case on the Domplatz (Cathedral Square). The banishment of private cars from the city centre had already been started before reunification, with some success. Now that a number of multi-storey car parks have been built, it is indeed possible to encourage the removal of motor vehicles from the townscape. Use of the pedestrian precincts for local public transport today is regarded as being beneficial to wheelchair users. This offsets their negative effect on the usability of the space.

Relationship of the city to the river
A topographical feature of the city is the river Gera flowing through it. Whereas the city used to have a great number of links with the river – the

Erphesfurt (Erphes Ford) through the river was where the first settlement began – its orientation toward the river has greatly reduced since the 1960s. The gangways at the Gera where people did their washing, and where crafts requiring water were performed, gradually fell into disuse. The houses along the Gera fell into disrepair; the city turned its back on the river. On the other hand, a park was laid out at Venice, a stretch of partly war-damaged land to the north of the *Krämerbrücke* (Krämer Bridge), but this had nothing to do with the old quarter's historic relationship to the river. With the redevelopment of the old quarter, the relationship of the city to the river is gradually being restored by the use of gangways, especially at points where houses are built directly alongside the river (Fig. 7.1). This is consistent with the Framework Plan which lays down that 'the waters of the River Gera and its vegetation system are to be managed and developed carefully and harmoniously as an elemental counterpoint to the "stonework of the city" ', without detailed directives being given. In any event, the arms of the river are to be made accessible to the visitor.

Expanses of green in the city

The Framework Plan also insists on there being greenery in the streets and courtyards of the city centre. Continuous stretches of 'green' are to be achieved in the city centre by joining up the public green spaces in the old quarter and in turn linking these with the major parks outside it to the south-west, e.g. the Nordpark to the north, and the grounds of the *Erfurter Gartenbau-Ausstellung* (Erfurt Horticultural Exhibition).

The old quarter as a residential and service area

The Framework Plan also stipulates that the city centre shall become the preferred location of the tertiary sector. The Plan is oriented toward putting

Fig. 7.1 The relationship to the river is restored step by step: new gangways to the back of the houses on the river Gera in the city centre.

the housing located outside the former city fortifications on ring roads bearing heavy traffic to other uses, so as to preserve the old quarter as a residential area. This is expected to have a positive effect on the structures worthy of conservation in the inner city, as by this approach the function mix of the buildings will be preserved. This mix of functions is postulated to create 'urban living'.

The fact that a disused industrial area, the Brühl, on the western edge of the old quarter, is available for functions requiring more space is a positive factor. Offices, theatres and hotels can be accommodated there without having to sacrifice the plot structure of the old quarter for this purpose – as in GDR times. The Plan stipulates that, as sites for small-scale commercial firms and service industries are typical of the old quarter, the plots are its backbone and give it atmosphere, and therefore should be promoted.

Local plans, project and development plans for segments

A further tool that can be deployed during redevelopment is the Local Plan, and at the beginning of the 1990s, project and development plans as well. Comprehensive instructions, even down to the design details, can be given under the Local Plan.

There are several Local Plans for the old quarter containing major, and often controversial, alterations. One related to the construction of an underground car park on the Petersberg, during the construction of which part of the south-east slope of the Petersberg was removed, then put back on top of the car park afterwards, making it scarcely visible from outside (Fig. 7.2). Another Local Plan related to the new Federal Labour Court building to the west of the former citadel. This modern, elegant building

Fig. 7.2 A new underground car park is located under the south-east part of Petersberg.

designed by the Berlin architect Gesine Weinmiller was opposed by the *Landesamt für Denkmalpflege* (Land Office for the Preservation of Historical Monuments) because it was to be built in the immediate vicinity of the Petersberg – where the said *Office* is accommodated in a former barracks (Fig. 7.3). Ultimately, however, it was executed in accordance with the prize-winning design without requiring any alteration of note.

In the redevelopment area itself, Local Plans are only drawn up where building work is to be carried out on a large scale. Under a 'Project and Development Plan', a tool specially introduced into the new *Ländern* after their accession to the Federal Republic of Germany to simplify the drawing up of the Local Plan, two department stores were built in the old quarter. This is a classic example of the confusion of the planning authorities in the years immediately following the East's accession to the Federal Republic of Germany. It is apparent from their façades that the design statute discussed above was not adhered to.

Competitions and expert advisory procedures

Whilst the purpose of the Federal Labour Court competition was not to upgrade the historical old quarter, several competitions for proposed projects have been held there. In 1991, a competition was held to develop the open spaces resulting from war damage and neglect to the north of the *Krämerbrücke* at *Am Venedig*. Since then, various competitions have been held, and more are planned, by private clients as well as public authorities, to close gaps and erect new buildings. In addition, the city also commissions so-called 'block concepts' for specific building projects.

Fig. 7.3 The new Federal Labour Court on the west of the former Petersberg citadel, built in 1999, was at first opposed by the Land Office for the Preservation of Historical Monuments, which is accommodated there in a former barracks.

Revitalization management for the old quarter

Redevelopment and revitalization

Redevelopment procedure

As is generally the case in German conservation, the purpose here is not to restore the authentic fabric that has been lost; instead one hopes to be able to remedy the scars left by destruction and neglect by providing a good, contemporary substitute. It is nevertheless customary today for redevelopment initially to mean conservation.

The redevelopment procedure does not confine itself to the drawing up of a plan. It also entails its implementation over a shorter or longer period of time. As part of the old quarter's redevelopment, five relatively small areas are being redesigned with the aid of separate redevelopment statutes under the 'full procedure' giving the municipal authorities wide-ranging powers of intervention. These were already designated redevelopment areas in 1991, before the whole of the old quarter became one. Covering 20 hectares, they make up approximately one tenth of the old quarter. This sector includes the *Andreasviertel*, a quarter which used to be the domain of the medieval city's craftsmen and which was in a thoroughly neglected and desolate state due to the proposal to build a four-lane carriageway through it. In the redevelopment context, the local authority has wide-ranging powers where all building questions in the redevelopment area are concerned, so as to be able to implement its redevelopment objectives. Not only do all building projects require a permit from the local authority, but all sales, encumbrances or division of plots of land do as well.

The city is entitled to require owners here to make a compensatory payment for any appreciation in the value of properties due to the area's redevelopment. Conversely, owners are not required to pay the development costs of renewing the infrastructure that are otherwise payable in this connection. These measures are intended to curb land prices and thus counter its exploitation for speculative purposes. Consequently, uses that generate less profit also have a chance in the redevelopment area.

Redevelopment agencies

The city uses two redevelopment agencies in the old quarter in accordance with Sections 157–159 *BauGB* (Building Code). These perform the tasks for the local authority, albeit under its supervision. It is for the city to define the objective and see that it is met. Redevelopment comprises, *inter alia*, measures relating to land organization, as well as repairs to buildings or premises which can only be carried out if the municipal authority purchases premises for the duration of the refurbishment or permanently. The redevelopment agency is responsible for acting as buyer and commissioning the restoration work, as well as performing the development work. Given the high degree of competence called for on the part of such a redevelopment agency, Section 158 *BauGB* (Building Code) requires that it be endorsed by the *Land* authority responsible. Essentially, the redevelopment agency performs the business functions associated with redevelopment. It is not allowed to deal with the planning of the area. This work is usually undertaken by independent planners, where it is not performed by the local authority itself.

In addition to the redevelopment agencies in the old quarter, there is another one in the *Oststadt*, an area of the city adjacent to the old quarter that developed since the *Gründerzeit* at the end of the nineteenth century. This is also a designated redevelopment area.

Redevelopment and repair financing

Financing is composed as a matter of principle of payments made by the owner and public sector grants. Assistance is given, on the one hand, for the city's functions in an area, so-called 'organizational measures'. These include not only development measures, but also, for example, the relocation of a troublesome factory out of an area. A second major focus of promotion is on building repair. Here it is the 'non-remunerative costs' that are subsidized, i.e. those which cannot be recouped when a property is put to an economic use after refurbishment. Funds are also provided for the provision of planning services in a redevelopment area.

The allocation of the funds, which come from *Land* and Federal resources, is the responsibility of the *Landesverwaltungsamt* which reaches a decision in concert with the local authority. It is only natural that, given their tight budgetary situation, the local authorities should first try to use the aid to cover functions of their own. A large portion of urban development aid is thus channelled into municipal projects in Erfurt. It is for the redevelopment agency, upon appointment, to administer the funds as the local authority's trustee; 20 million DM annually in Erfurt.

Preservation of historical urban monuments

Erfurt receives financial assistance as part of the Federal Programme entitled *Städtebaulicher Denkmalschutz* (Preservation of Historical Urban Monuments). This aid programme was launched for the cities in the east of Germany after reunification in 1991 and, from the very outset, was attended by a group of experts appointed by the *Bundesbauministerium* (Federal Building Ministry) to advise the cities on their planning. At the same time, a Federal competition was held in which Erfurt received a gold medal in 1994. On the one hand, this competition had a positive effect by publicizing redevelopment; on the other, it served as a stimulus for increased redevelopment activity in some places, or at least for publicizing it. The Preservation of Historical Urban Monuments Programme enables local authorities to support the above projects, whilst limiting their contribution to a mere 20 per cent of the costs (only 10 per cent in Thüringia).

Individual projects as part of redevelopment

Major projects implemented in Erfurt's old quarter under the Preservation of Historical Urban Monuments Programme include restoration of the *Collegium maius*, the former main building of Erfurt University which was destroyed during the war. The building is a reconstruction that presents itself as one 'strictly modelled' on a Gothic style of architecture that it possibly once had (Schleiff, 1999) – a not uncontroversial approach that is agreeable in terms of healing the wounds in the old quarter, irrespective of the details of its execution (Fig. 7.4).

The *Kulturhof Zum Güldenen Krönbacken* is a project already completed. At the beginning of the 1990s, a municipal cultural centre was created here with the aid of the Preservation of Historical Urban Monuments Programme by restoring the medieval building fabric and adding a modern extension. This is an example of putting courtyards to use, as called for in the Framework Plan (Fig. 7.5).

In addition to these municipal properties, there are other properties that have been put to mixed use as residential accommodation, and for trade and commerce. In *Marktstraße* alone there are three complexes in which the courtyards have also been integrated. All three are seventeenth-century buildings that were already falling into decay.

As well as some smaller buildings, one of which won the Thüringian Prize for Architecture in 1999, some quite large building complexes are also being erected in the old quarter in place of buildings already demolished or beyond repair. In this way, numerous apartments are being built in *Michaelisstraße* as part of a municipal housing association project. Admittedly, valuable conservation material was also demolished in the course of this building project.

Fig. 7.4 Michaelisstraße in Erfurt: in the middle of the picture is the rebuilt Collegium maius (destroyed in 1944), 'modelled strictly' on a Gothic style of architecture that it possibly once had. (Schleiff, 1999.)

Fig. 7.5 Now a public space: courtyard of the house Zum Güldenen Krönbacken. Left: the former 'Waidspeicher' (store house for 'Waid', a plant used in the dying process), now a gallery; right: the new utility building of the gallery, built in 1994.

The churches as owners of historical monuments

The main owners of historical monuments in the old quarter, in addition to the city, must be the Evangelical and Catholic Churches, who own some twenty churches and numerous other buildings as well. In the Middle Ages Erfurt had several churches, but during the Reformation quite a number of these were demolished; in some cases the steeples have survived. Today, the church authorities are scarcely in a position to maintain these buildings in conservation terms out of their own resources, and consequently they are dependent on subsidies from various sources.

Societies

The oldest society to concern itself with Erfurt's architectural history is the *Bürgerinitiative Altstadtentwicklung* (Civic Action Group for the Development of the Old Quarter). It was founded in 1987 to oppose the demolition of the *Andreasviertel* and it survived the demise of the GDR. This Action Group is continually launching campaigns to draw the public's attention to the cultural heritage. Its purpose is not only to prevent irreparable damage being done – like the demolition of Erfurt railway station in 1888/94 – but to prompt debate about how the old quarter should be developed. Its latest venture is a project to reopen old alleys and paths. Twenty-nine of these paths have been listed, and whilst most of them still exist, they are currently not generally public thoroughfares. They include five footbridges that used to lead over the Gera and which, if reinstated, would increase awareness of the old quarter's relationship to the river.

In the meantime a new society has been set up in the *Andreasviertel* to try to save the barns still found there. The *Erfurter Geschichtsverein* (Erfurt Historical Society) also helps to preserve the old quarter by annually publishing material on the architectural heritage.

Foundations

The work of various foundations is also of major importance for the old quarter's architectural heritage: the *Deutsche Stiftung Denkmalschutz* (German Foundation for the Protection of Historical Monuments) gives financial support to various projects in Erfurt. These currently include a granary in *Michaelisstraße* and the *Krämerbrückenstiftung* (Krämer Bridge Foundation).

The *Krämerbrückenstiftung* was founded as a trustee of the city to ensure conservation of the Krämerbrücke and the buildings on it. Most of the houses on the bridge – all but three – belong to the city. The Foundation's task is to see that they are put to suitable use by ensuring a mix of residential accommodation, handicrafts, trade and commerce. It is also required to ensure redevelopment and act as a recipient of private-sector donations.

The *Stiftung Thüringer Schlösser und Gärten* (Foundation for Thüringian Palaces and Gardens) owns the former monastery church of St Peter which has served as a warehouse since the nineteenth century. The Foundation was set up by the *Land* in order to be able to place properties of particular cultural importance in its care. The Foundation is responsible for the preservation, running and use of the properties entrusted to it. It is funded by the State of Thuringia.

In recent years the *Deutsche Bundesstiftung Umwelt* (German Federal Environment Foundation) has also contributed considerable sums of money to projects to repair environmental damage to buildings such as the former church of St Peter and the Cathedral.

Municipal and* Land *urban renewal authorities

Since 1997, the city's urban renewal functions and the permit procedure under the Thüringian Law on the Protection of Historical Monuments have been concentrated in an *Amt für Stadterneuerung und Denkmalpflege* (Office for Urban Renewal and Conservation). The function of this inter-disciplinary Office is to draw up fundamental urban renewal objectives and the plans based on these, co-ordinate redevelopment as a whole and issue the redevelopment and conservation permits required in this connection. This Office co-operates with the competent department in the *Landesverwaltungsamt* (Land Administrative Office) in urban redevelopment matters. In order to perform its duties under the Thüringian Law on the Protection of Historical Monuments, which the city can only do in concert with the *Landesamt für Denkmalpflege* (Land Office for the Preservation of Historical Monuments) and the *Landesamt für Archäologische Denkmalpflege* (Land Office for the Preservation of Archaelogical Monuments), the city co-operates closely and fruitfully with these. The legal procedure has already been described in a corresponding publication (Pickard, 2001).

Erfurt works in concert with the two *Land* Offices in such a way that there are no differences of opinion between the municipal and *Land* authorities (Wittich, 1999). In practice, this consensus principle means, of course, that economic considerations are taken into account without lengthy procedures being required. This may be good for economic development, but it is not always conducive to the reputation in particular of the *Landesamt für Denkmalpflege* as a conservation authority.

In addition to the above municipal inter-disciplinary office within the Building Department, other sections of this Department, such as the Office for Transport and Town Planning, are important. Other relevant departments include the Mayor's Department (Office for Urban Development, Statistics and Elections, responsible for urban development planning), and the cultural directorate in the Culture Department is also responsible for preservation of the old quarter.

Commercial exploitation of plots in the old quarter

It is not just in this century that plots of land have been joined together, as we see from the *Krämerbrücke*, the stone arched bridge over the Gera with its houses, where the number of plots was halved in the last century. Even so, the small-sized plot is still far more in evidence in Erfurt than in comparable cities. Now the problem is not solved by façade cosmetics as attempted by the new C&A building in Erfurt's *Schlösserstraße* with its façade divided up into individual structural elements, even if this meets the requirements of the design statute (see above) (Fig. 7.6). In addition, a third storey was added to a medieval house incorporated here. By contrast, given the small-scale housing units of the *Andreasviertel*, it would not have been necessary to build quite large units, even though this must be qualified by saying that the *Moritzhof*, a sizeable block of buildings containing 114 council flats, had already been built in this area in 1920/21. What does seem possible and in the spirit of the Framework Plan are individual buildings like the house at *Glockengasse* 32 (Architect Oßmann), which recently won the Thüringian Prize for Architecture, or the reconstruction of the

Fig. 7.6 The façade of the department store on the right is an example of the application of the design statute and 'façade cosmetics'.

Ziegengasse where the buildings are completely modern but modelled on the earlier structures (Architect Rabe).

City management in the inner city

Erfurt does not have a 'City Manager'. Such functions are performed by the *Verein Citymanagement e. V.* (City Management Society). Anyone with a legitimate interest in the city's development can become a member. In 1997, the Society had twenty-three founding members, and today it has a membership of sixty.

Its chief concern is to steel the inner city against the commercial competition outside the city limits. This includes gathering and disseminating all the relevant information and making it available to members in an 'Info Letter'. The Society seeks to co-ordinate activities in the inner city, and in this does not confine itself to commerce and services. The Society co-operates closely with the authorities which also consult with it on questions of inner city design.

Environmental management and traffic

With the collapse of the GDR, the environmental situation in the inner city gradually improved. As Erfurt is situated in a valley, the burning of heavily sulphurous coal in winter up to 1989 often made conditions intolerable, and the negative effect on buildings was appreciable. Now that natural gas is the chief fuel used and industry has largely disappeared from Erfurt, the situation has improved considerably. Today's air pollution is largely due to motor traffic which has increased enormously since 1989.

Traffic in the old quarter of Erfurt has been restricted for decades. The main street complexes – *Anger, Bahnhofstraße* and *Marktstraße* – are almost entirely pedestrian precincts. Private motorists are not allowed to drive through the old quarter. In recent years, high-rise car parks have been built on the approaches to the old quarter. This has made the inner city more accessible to private vehicles and made it more competitive with the shopping centres on the outskirts of the city. The *Stadtwerke Erfurt Parken GmbH* (Erfurt Public Utility Parking Ltd), owned by the city, runs the high-rise car park at the *Domplatz* (Cathedral Square) and endeavours to make it – and accordingly the old quarter – more attractive by giving concerts there, for example. Bicycles can also be rented there. The inner city also profits from the fact that all the tram and local bus routes intersect and have their main connecting stops there. In the old quarter, trams provide public transport in the pedestrian precincts as well.

The railway is of the utmost importance for local public transport, the main station being situated on the edge of the old quarter. The city is to be linked up with the railway's ICE (Inter-City Express) network, a proposal currently called into question following Federal Government decisions. The railway authority considered demolishing the main station's historic nineteenth-century *Inselgebäude* (island building), surrounded by railway tracks (Fig. 7.7), as a necessity for the planned ICE link-up. It is the last surveyed example of a special type of railway station in Germany. The

Fig. 7.7 The Inselgebäude at Erfurt Central Station, surrounded by platforms, is the last surviving example of a special type of railway station in Germany. In keeping with the ICE link-up, a new railway station is planned and the old one is to be demolished.

main building is reached by a tunnel and stairs from the entrance hall, situated on the square in front of the station. The municipal and *Land* conservation authorities approved this questionable demolition request. The Custodian of Monuments of the day felt faced with a choice between the ICE and a station designated as a historic monument, and he opted for the ICE for the good of the city and the *Land*. Alternative courses were not considered, even though alternative plans existed.

Tourism and heritage management

Planning and tourism

The urban planning authorities consider a revitalized old quarter a decisive geographical factor in attracting jobs to the city, as well as being a tourist attraction. One of the objectives of the Framework Plan is to develop the Petersberg. With this in mind, the seventeenth-century citadel on the north-west edge of the old quarter has been undergoing restoration for years. Admittedly, this also means that parts of structures long since been lost, such as sizeable sections of the wall and a whole series of sentry posts and are being reconstructed on a grand scale. The various structures are being restored according to their different construction periods, and so ultimately a quasi-historical site will be created that never existed in this form.

Tourism marketing

In order to improve the city's tourism marketing, the city transformed the municipal 'Tourist Information' into the 'Tourist GmbH Erfurt' in 1998. The Tourist GmbH's functions include tourist services such as hotel reservations, guided tours, package tours and organizing congresses. Its services for the city entail, above all, advising on tourism, there being development potential in closer links between the tourist industry and urban development planners.

Tourist GmbH analysed the tourism situation last year and found that 7 million visitors come to the city every year, 500,000 of them on business. Erfurt hotels record 500,000 overnight stays per annum, whilst there are 2.3 million private overnight stays, approximately 10 per resident per year. The main tourist attractions are the Cathedral and the *Krämerbrücke*.

In January 1999 Erfurt joined the 'Historic Towns of Germany', an alliance of thirteen cities of historic importance including, *inter alia*, Augsburg, Bonn, Freiburg, Heidelberg, Lübeck and Regensburg.

Naturally, Tourist GmbH and the municipal authority are on the Internet, but they also employ a range of advertising tools to generate public interest in the city. In future, Tourist GmbH is to be in charge of all city marketing.

The problem of too much tourism that the local people find irksome in other cities is not (yet) one for Erfurt.

Festivities and functions as attractions in the old quarter

A whole range of events attracts public attention to the old quarter. First and foremost, there are the fairs on the *Domplatz*, in particular the Christmas Market which has meanwhile established itself nationwide. At the end of spring, the '*Entenrennen*' (Duck Race), in which plastic ducks are raced on the river Gera through the old quarter, is an attraction for families in particular. The *Krämerbrückenfest* (Krämer Bridge Festival) at the beginning of summer attracts thousands of spectators every year.

In late summer the theatre holds a *Domstufenfestival* (Cathedral Steps Festival) in which the well-known steps of the Cathedral and the Severi Church serve as a backdrop and stage. Since 1993 the 'European Open Monument Day' has been celebrated in Erfurt too. As of 1997 there has been a whole week of festivities under a special theme in addition. This year the theme was 'Wege' (Paths) and drew attention to the pathways described above. Organization of inner city events is the province of the urban culture division of the municipal authorities. It is responsible for 'culture management' and is in charge of all municipal cultural facilities – including all the city museums located in the old quarter – as well as art promotion, socio-culture, events and markets.

Conclusions: sustainability of urban renewal

Sustainability – even if not constantly stated explicitly – is a major objective of old quarter conservation. Admittedly, people scarcely ever ask whether today's conservation of the old quarter will also meet future development requirements. It goes without saying, however, that preservation and renewal of existing structures are easier on the environment than their complete replacement. There is no doubt that the tourist potential and proposed mix of uses planned and gradually being implemented in the old quarter present a great opportunity for a future development of the city for coming generations as well. This presupposes that we heed the words of

John Ruskin, which date from the middle of the last century, who said of historic buildings: 'They are not ours. They belong partly to those who built them and partly to all the generations of mankind who are to follow us . . .' (Ruskin, 1849). In Erfurt conditions for adopting this philosophy are now favourable.

References

Altwasser, E. (1999): 'Unter der Stadt – Erfassung der Kelleranlagen in Erfurt', in *Heimat Thüringen* 6 (2/3): 19.

BauGB (1997): 'Baugesetzbuch', in: *Bundesgesetzblatt* I: 2141.

Brüggemann, S. and Schwarzkopf, C. (1997): *Carl-August-Allee Weimar*, Weimar.

FNP (1996): *Landeshauptstadt Erfurt, Stadtverwaltung: Frühzeiti Bürgerbeteiligung zum Vorentwurf – Flächennutzungsplan,* Erfurt (map with explanations).

FNP (1999): Landeshauptstadt Erfurt, Stadtverwaltung: *Flächennutzungsplan –Entwurf – Öffentliche Auslegung,* Erfurt 1999 (map with explanations).

Imhof, M. (1993): *Bauen und Wohnen in einer fränkischen Kleinstadt vom 16. Bis 19. Jahrhundert am Beispiel von Königsberg in Bayern*, Bamberg, p. 11, footnote 26.

Pickard, R.D. (ed.) (2001): *Policy and Law in Heritage Conservation,* Conservation of the European Built Heritage Series, Spon Press, London (See Chapter 7: Germany by Brüggemann, S. and Schwarzkopf, C.).

Reinhardt, H. (1999): 'Mittelalterliche Wohnbauten in Thüringer Städten', in *Heimat Thüringen* 6 (2/3), 18.

Ruskin, J. (1849): *The Seven Lamps of Architecture*, 'The Lamp of Memory', XVIII.

Schleiff, H. (1999): 'Zum Wiederaufbau des Collegium maius', in *Stadt und Geschichte* 3, Erfurt.

Wittich, U. (1999): Notes of an interview with Mr U. Wittich, Town Administration of Erfurt, by Brüggemann, S. and Schwarzkopf, C.

Other sources of information

Homepage: *www.erfurt.de,* 'Sanierung in Erfurt'

Amtsblatt 24.6.1992 *Amtsblatt der Stadt Erfurt*, Erfurt

Amtsblatt 25.11.1992 *Amtsblatt der Stadt Erfurt*, Erfurt

Elene Negussie

Dublin, Ireland

Introduction

This chapter explores the Irish experience of historic and architectural heritage management as reflected in the city of Dublin, a city that has, over the past decade, experienced considerable renewal and regeneration of inner city areas concurrent with rapid national economic growth. A brief historical background to the evolution and character of the problems that conservationists, local and national authorities have had to deal is followed by an examination of issues that have had major impacts upon the built heritage: inner city restructuring and decline, property markets, traffic management, legislation and cultural attitudes.

Although Dublin can trace its origins back over a thousand years, few elements of the townscape pre-date 1650. Economic prosperity and political importance during the eighteenth century endowed the city with a heritage of fine public and domestic buildings. Subsequently, economic depression was to ensure that this city was to become preserved almost intact until the middle of the twentieth century, when Walter Bor (1967: 293) commented portentously:

> Dublin is a city of quite exceptional character and beauty, with a long history and vibrant cultural tradition. The closely interwoven pattern of eighteenth century streets with their remarkable consistency of scale and material, once quite a common sight in Europe, can now be found only in Dublin, and the city's architectural heritage becomes rarer and more vulnerable every day… Dublin's past growth has been relatively slow and steady compared with other cities. The city has not been radically altered in character by the effects of twentieth century development. The predominant part of the eighteenth and nineteenth-century fabric of the city has survived and the basic pattern of scale and material remains intact. Thus on the one hand this city has the

great inheritance of Georgian domestic architecture of a unique quality and scale, but on the other hand there is an acute threat that the effects of rapid expansion over the next two decades could be all the more damaging.

A number of factors conspired to place that valuable heritage at threat: economic growth and the profitability criteria of commercial property developers, cultural attitudes towards the Georgian building stock, in addition to the activities of the public sector itself.

Until relatively recently, little attention was paid to conservation except for the care of certain set-piece public buildings and sporadic efforts by conservationists. As Maurice Craig opines, 'the survival, almost intact, of Dublin down to, say, 1950 was a freakish circumstance. It was directly due to our relative poverty in the twentieth century, coupled with the great wealth of the city, if not the country, in the eighteenth' (Craig, 1975, 13). Thus, the preserved character of Dublin resulted rather from the negligible pressures that would have been occasioned by economic development rather than from deliberate attempts to protect the architectural heritage. Although there have been major changes in attitudes towards the built heritage, especially during the last decade (Negussie, 1996), cultural attitudes towards the built heritage have been a major problem in the past. After the Irish Free State was formed in 1922, rural areas long remained the priority and there was a negative attitude towards the conservation of Dublin's 'Georgian' and 'Victorian' heritage, as it reflected the period of subservience to British rule. The passage of time has encouraged a revaluation of that heritage and recognition that these buildings, having been built by Irish craftsmen, comprise an essential element of the national patrimony.

Attitudes amongst professions dealing with the built environment were strongly biased against historic buildings. Not only was there a desire to promote new architectural styles as evidence of the changing image of the 'New Ireland', but negative attitudes towards the historic stock were hardened by its associations. In many areas of the city, Georgian buildings had become synonymous with the slum tenements that characterized Dublin's economic depression throughout the nineteenth and the first half of the twentieth centuries. During the nineteenth century, as middle-class Dubliners moved out to more commodious new suburbs, a considerable proportion of Dublin's eighteenth-century housing stock was tenemented, particularly to the north of the river Liffey. These were often divided into single-room dwellings, each of which would house a family, perhaps with lodgers to assist in the payment of rent. Some housing, developed on short leases, had been poorly built originally and, by the late nineteenth century, had become dangerous and prone to spontaneous collapse, sometimes with loss of life. Poverty, overcrowding, insanitary conditions and elevated death rates that exceeded any British city of the late nineteenth century, pointed to a serious tenement housing problem which the newly independent state was obliged to face. Without consideration of the architectural heritage, the local authority (Dublin Corporation) dealt with the 'tenement problem' by clearance.

The espousal of the modern movement in architecture together with a profound prejudice against the 'Georgian' townscape was to have devastating effects on the urban landscape (Nowlan, 1982).

Economic growth after 1960 generated a demand for office space in the city. The stock of eighteenth- and early nineteenth-century buildings, set out around squares and along wide thoroughfares, underwent conversion from residential to office functions. Regrettably, and all too frequently, inadequate attention was paid to the protection of valuable interior items – even by public sector users. Marble fireplaces were ripped out and cornices irreparably damaged. Until a decade ago, there was no official protection of interiors. For a city which was reputed as possessing some of the finest domestic plasterwork in Europe (Rowan, 1980), the result was little short of tragic.

Economic expansion also generated an office development boom. Unfortunately, 'the vastly increased demand and competition for land area for office use created an economic situation in which the value of cleared sites within the central urban area, particularly Georgian districts, was often greater than that of the buildings which originally existed on the site' (Kearns: 1983, 60). During the following decades, the major focus of redevelopment activity lay where the greatest potential development profit was to be made. This was located in the prestigious south-east inner city of Dublin 2, where the quality of the remaining Georgian housing was at its best. It gave rise to widespread demolition of eighteenth- and nineteenth-century buildings, replacing them with space-efficient schemes built in a modern idiom or as pastiche replicas.

Two contrasting scenarios therefore developed. To the north of the river Liffey, limited user-demand of any type whatsoever resulted in much of the building stock falling into decay. As a result, historic buildings in some of Dublin's most important Georgian areas became exposed to neglect and decay, resulting in dereliction. Fig. 8.1 portrays a Georgian doorway in Henrietta Street, where historical buildings have been allowed to fall into decay, due to its commercially disadvantageous location in the north inner-city. To the south of the river, the building stock became threatened by conversion to offices or redevelopment and numerous buildings were either altered or demolished to this end (MacLaran, 1993). Fig. 8.2 depicts Georgian terraced houses on Merrion Square, a commercial and prestigious headquarter in the south inner city. The square is today one of the top-priority conservation areas in Dublin.

Another major factor threatening the stock of historic buildings in the city was the wide-scale plans of the city's roads engineers to improve traffic flows through road-widening schemes. The Dublin Development Plan of 1980, proposed that 82 streets should be widened in the inner-city alone (McDonald, 1989) and the Corporation had engaged over a number of years in site assembly to this end. By the late 1980s, Dublin Corporation was the owner of a considerable stock of derelict property which was earmarked for demolition to implement road-widening and also for housing, parks and community uses (MacLaren, 1993) (see Fig. 8.3).

Fig. 8.1 Georgian doorway in Henritta Street (north inner city).

Fig. 8.2 Terraced Georgian houses on the prestigious Merrion Square (south inner city).

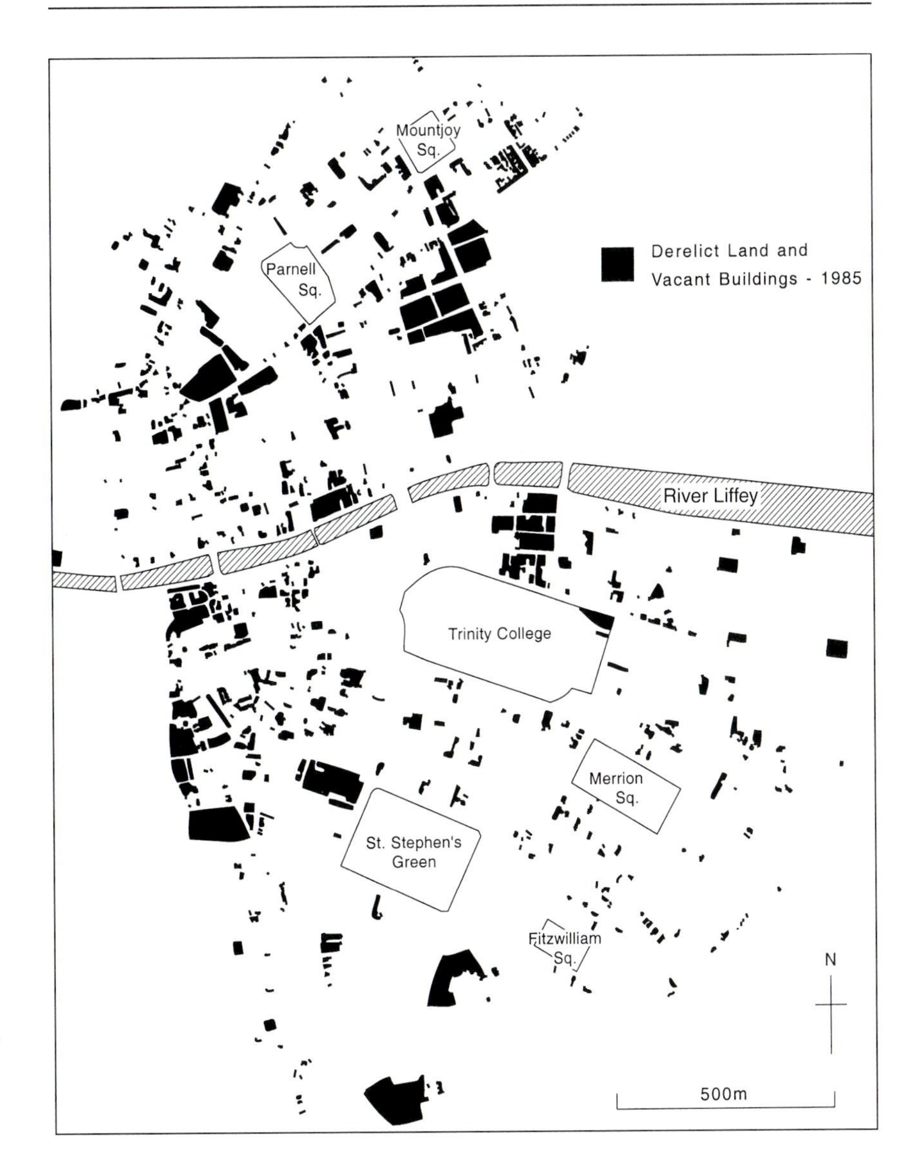

Fig. 8.3 The wide-scale existence of derelict land and vacant buildings in central Dublin in 1985 (Source: MacLaran, 1993.)

Action by conservation groups

Alongside the widespread demolition and neglect of historic buildings in Dublin's inner city was a growing concern by the conservation lobby. Conservationists played a significant role in raising awareness of the built heritage in Ireland and, as a result of the many urban conservation battles, they began to receive growing public attention in the 1960s and 1970s.

The Irish Georgian Society came to play an important role in one particular controversy that involved the demolition of a Georgian terrace of sixteen houses in Fitzwilliam Street. The buildings, which had been acquired by the Electric Supply Board for office-use, were demolished following a public enquiry and replaced by a modern structure. Fig. 8.4 illustrates the modern office block with remaining Georgian houses.

Fig. 8.4 The replacement of a Georgian terrace with a modern office block in Fitzwilliam Street.

An Taisce, the National Trust for Ireland, experienced an expansion of membership throughout the 1960s and has become the largest environmental organization in Ireland. In addition, it was designated as a Prescribed Body under the Local Government Act, 1963, giving recognition of the organization in the planning process through its right to be consulted over planning and development issues (Mawhinney, 1989). The Dublin Civic Trust is another conservation group that has been directly involved in conservation of the built heritage. Student groups have also played a significant role through lobbying and the occupation of historic buildings under threat of demolition.

The Dublin Crisis Conference, held in 1986, brought together a wide range of conservation, community and environmental groups and called on the Government and Dublin's local authorities to recognize that the city was in crisis. A 'Manifesto for the City' was published the following year, recommending that the Government should increase its spending in public transport, criticizing Dublin Corporation's road-widening schemes and emphasizing the need to bring new life into the inner city through a sensitive approach to urban renewal (McDonald, 1989).

Listing

Under the Local Government (Planning and Development) Act, 1963, Ireland received its first effective modern planning legislation. However, under this legislation, the demolition of buildings required no permission unless they had been listed for preservation or protection in a Development Plan. The Housing Act, 1969, later extended protection to all residential buildings. The first lists of buildings to be preserved or protected appear in the 1971 Development Plan. These lists were subsequently amended and expanded in later Plans. However, although the Dublin City Development Plan, 1980, listed 613 houses for 'preservation', only 23 of these were situated north of the river Liffey (An Taisce, 1985b).

Even buildings listed for protection in Dublin remained at risk. Between 1980 and 1985, some 80 listed buildings had been either demolished or permission had been given for their demolition. Another 50 had suffered unauthorized material alterations detrimental to their architectural character (An Taisce, 1985b). The 1991 City Development Plan substantially increased the number of listed buildings by reference to four different lists:

List 1 (for preservation)	206
List 2 (for protection)	726
List 3 (state owned buildings)	33
List 4 (interior features)	101

Despite this improved situation the regime of protection remained weak due to the inability of the planning system to safeguard buildings. The system of protection in Ireland remained on a basis of local determination by 'lists' being created through the objectives of development plans. There was no clear form of statutory procedure on a national basis to safeguard the architectural heritage as demanded by the Granada Convention (Pickard, 1998). Moreover, in Dublin in particular, a lack of building maintenance codes and the absence of penalties for owners that demolished or neglected historic buildings long thwarted any serious attempt to conserve the historic core. Indeed, the easiest way historically for a property owner wishing to divest himself of the burden of an architecturally important building, on a potential redevelopment site, was to let it fall into decay. In some cases, decay was assisted by the removal of valley gutters to facilitate the penetration of the structure by water. Almost inevitably, inspectors from the dangerous buildings section of Dublin Corporation would then condemn the structure as dangerous and require its demolition (McDonald, 1989).

Towards conservation

In 1985, the Irish Government signed the Granada Convention, accepting the concept of 'integrated conservation' through which conservation should be considered not as a marginal issue, but as a major objective of town and country planning. Whilst the convention advocated measures to strengthen legislation and the administration of architectural heritage protection, it was only in 1997 that the Government ratified its articles. Moreover, it is only recently that measures have been taken to review the system in operation (Pickard, 1998) and to enact comprehensive legislation on conservation (MacRory and Kirwan, 2001).

The emerging policy and planning framework: property-led regeneration

Since the 1930s, like many other European cities, Dublin experienced an almost continuous reduction in its inner city resident population (Horner, 1985). Slum clearance, the decanting of inner-city communities to peripheral housing estates and the more voluntary relocation of middle-class residents to suburban owner-occupied areas resulted in the inner-city resident population falling from the 1960s while, at the same time, the wider metropolitan population grew to over 1 million. During the 1970s, an inner-city economic crisis was created by a sharp reduction in industrial employment associated with the relocation of activities to suburban industrial estates. Inner-city unemployment was further exacerbated by changes in cargo handling in the port, with unemployment rates exceeding 35 per cent in inner-city areas. Simultaneously, the service sector increased its share of employment, resulting in even greater demands for office space and displaced industrial activities (MacLaran, 1993). It became imperative for the Government to address this inner-city crisis and it attempted to do so in the mid-1980s, through the vehicle of property-led regeneration.

Government policy on urban renewal

In the mid-1980s, an Urban Renewal Programme was formally introduced by the Government in response to the urgent need for the revitalization of inner-city areas that had fallen into physical, social and economic decay. The Programme targeted the cities of Dublin, Cork, Limerick, Waterford and Galway. The Urban Renewal Act, 1986, empowered the Minister of the Environment to designate urban areas in need of renewal and to allow special tax relief to encourage private sector-led redevelopment in such areas. The Finance Act, of the same year, allowed special remissions of rates including capital allowances for commercial development, subsidies to occupiers by means of rates remissions and additional rent allowances for commercial activities, and special income tax allowances for residential owner occupiers (An Foras Forbartha, 1986).

The Urban Renewal Act of 1986 also provided for the establishment of an independent development body, the Custom House Docks Development Authority, to be in charge of the development of the designated Customs House Docks. Other areas to be designated in Dublin under this Act were tracts in the north inner city, along the north and south quays of the river Liffey and around Christchurch Cathedral at the heart of medieval Dublin (Dublin Corporation, 1986). The areas were later expanded to include wider tracts of the inner-city area that were in need of renewal (Fig. 8.5).

However, the thrust of the incentives often had a negative impact on the built heritage as they tended to be geared towards new construction rather than the rehabilitation of older buildings. The incentives were introduced at a time when the property development sector was about to enter a phase of intense development activity in response to rapid economic growth. The poorly developed legal framework for conservation contributed to the

Fig. 8.5 Designated areas for urban renewal in Dublin's inner city, 1986–94 (Source: MacLaran, 1996).

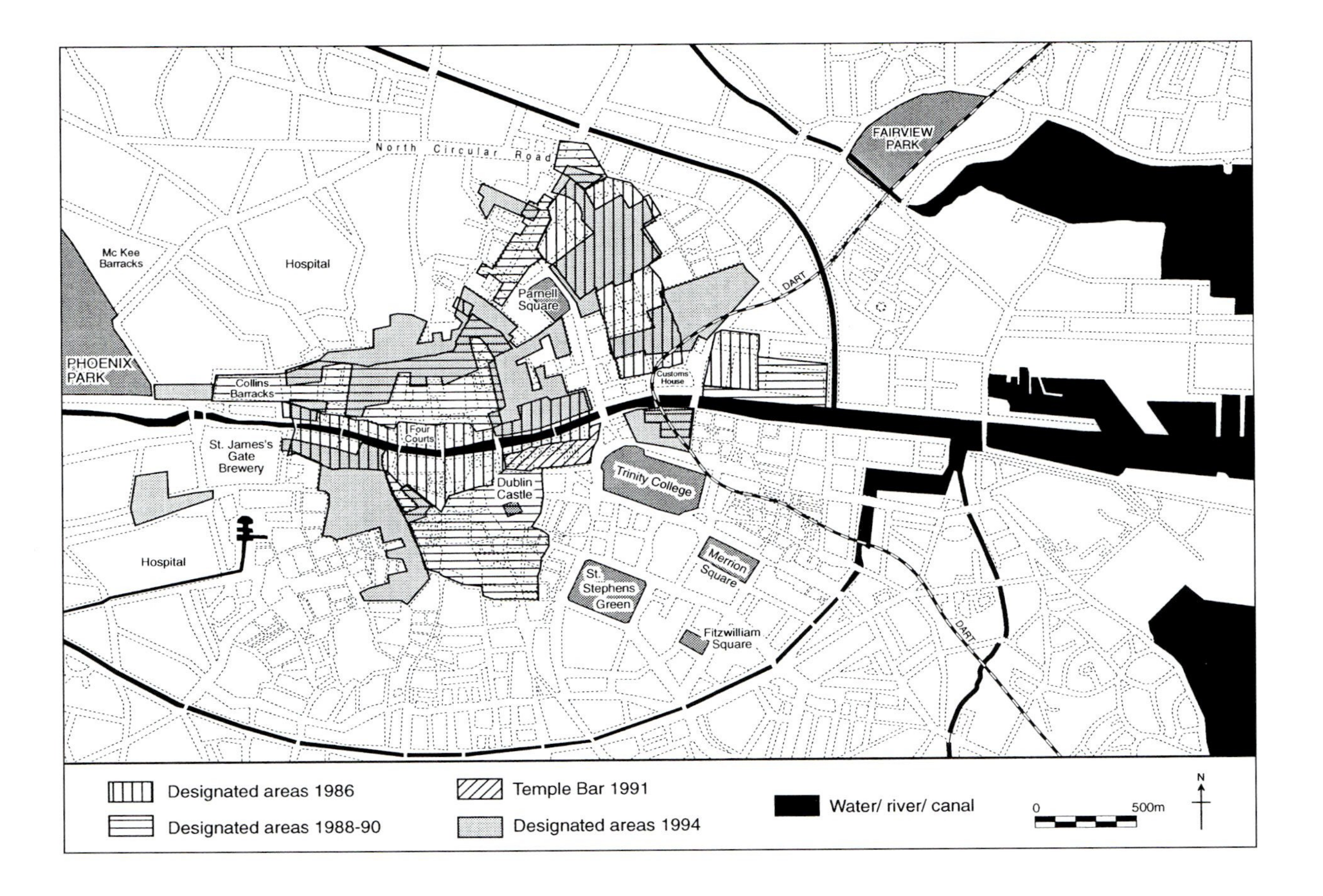

creation of major difficulties, particularly in secondary areas that became an important focus for office development. Only belatedly was there any attempt to reorientate the thrust of the incentives towards the encouragement of renovation.

In 1996, a consultant Study on the Urban Renewal Schemes was commissioned by the Department of the Environment to evaluate the results of the Urban Renewal Programme. The study found that whilst it had proved successful in attracting private sector investment into areas of physical decline, it had in many cases been unsuccessful in terms of architecture, design, conservation and social aspects (KPMG, 1996). The study concluded that there was a need for a more integrated approach to urban renewal. In 1998, the Government responded by the introduction of a new programme of Integrated Area Plans to address physical, economic, environmental and social aspects (DoE, 1997a). In Dublin, there now exist five such plans: the North East Inner City; O'Connell Street; Historic Area Rejuvenation Project (HARP); the Liberties/Coombe; and Kilmainham/Inchicore. The aim of the local authority is to achieve conservation within these areas through local plans developed in co-operation with local communities and environmental groups (Dublin Corporation, 1999).

The role of national and local authorities and agencies

The planning system in Ireland is a centralized system with national government enacting legislation and planning regulations and local government being responsible for the production of development plans and for deciding on planning applications for permission to develop. A politically independent planning appeals board, *An Bord Pleanála*, forms part of the planning system to which planning decisions can be appealed (Newman and Thornley, 1996). A unique feature of the Irish planning system is the legal right of third parties to appeal decisions, thereby providing opportunities for environmental lobby groups to intervene in the planning process (MacLaran, 1993).

The main responsibility for developing and implementing policy on the urban built heritage lies with the Department of the Environment and Local Government and the Department of Arts, Heritage, Gaeltacht and the Islands (DACG). The former is responsible for policy on the urban environment through the local government system, e.g. urban renewal, physical planning and environmental protection. This is partly funded by the European Union assisted Urban and Village Renewal Programme (DoE, 1994). With new legislation coming into force shortly, it will directly fund local authorities towards urban conservation. The latter department is responsible, through the Heritage Service (*Dúchas*), for the National Inventory of Architectural Heritage and for the management of national monuments in state care. It also has a shared advisory role with the Arts Council and the Heritage Council on planning issues which relate to buildings of artistic, architectural or historical interest (DACG, 1996).

The two government departments have jointly developed a new legislative framework for conservation of the built heritage that will have particular relevance in Dublin. With the introduction of new legislation, local authorities will acquire new responsibilities for the protection of the built heritage, for example – mandatory listing and enforcement of conservation. This will be supported by the Heritage Council (an independent body responsible for promoting the national heritage including the architectural heritage by statute), which will develop a framework for further integration of local development plans including a proposed extended grant giving function (Kirwan, 1998; MacRory and Kirwan, 2001).

Development agencies and authorities are concerned mainly with regeneration of inner-city areas. However, they have developed an important role as they can exert direct or indirect influence over the way in which the historic built environment is shaped in designated urban renewal areas. For example, the Dublin Docklands Development Authority is an independent Special Purpose Authority, responsible for overseeing the redevelopment of some 526 hectares of redundant docklands. The area contains a built heritage that reflects the city's industrial and trading history, including several important set-piece buildings such as the classical Custom House. As will be discussed later, Temple Bar Properties Limited and Temple Bar Renewal Limited have also played a significant role in the renewal of Temple Bar, a zone of mixed low-grade commercial activity lying between the city's two main shopping streets just south of the river Liffey. While the former organization is a quasi-state development company acting on behalf of the Government, the latter was set up as a vetting and approval agency to ensure that subsidies would be directed towards only a prescribed range of activities in Temple Bar. Such agencies will be required to have a more conservation-orientated approach in the future.

The recording of the built heritage in Dublin

The recording of the built heritage is characterized by a fragmented approach and no overall inventory has been completed to date. It remains a major task of local and national authorities to carry out inventories of the building stock in Ireland. As a response to the Granada Convention of 1985, the Office of Public Works established a National Inventory of Architecture in 1991. The aim was to achieve a standard of consistency for the recording of Ireland's post-1700 architectural heritage, complementing the Archaeological Survey of Ireland. The inventory, which has to date concentrated on towns other than Dublin, is undertaken by the Heritage Service and operates under the Department of Arts, Heritage, Gaeltacht and the Islands.

However, in the absence of an overall recording strategy for Dublin, the voluntary sector has played an important role. The *Urbana* report was one of the first attempts towards systematic identification of historic buildings covering all buildings under List One in the city development plan (An Taisce, 1982). In 1985, An Taisce also produced an inventory of the Temple Bar area, noting significant groups of buildings and examples of important paving sequences and stonework features (An Taisce, 1985a). In the mid-

1980s, the Irish Architectural Archive also undertook a survey of interiors, some 100 of which were subsequently considered sufficiently important to become listed by Dublin Corporation for protection.

In the mid-1990s, a Dublin Environmental Inventory was established as an outcome of a European Union assisted initiative by Dublin Chamber of Commerce. A Built Environment Survey, undertaken by the School of Architecture at University College Dublin (UCD), formed part of this. The format of this inventory was consistent with the National Inventory of the Architectural Heritage and adapted to the needs of Dublin. Furthermore, it was prepared in accordance with the requirements of the Core Data Index which are recommendations on inventory and documentation methods adopted by the Council of Europe. The survey was carried out in the inner-city and included building and street inventories (Kealy and O'Rourke, 1995). The project resulted in a complete inventory of all List One buildings and all List Four interiors appearing in the Dublin City Development Plan, 1991. The initial objective was to create a Geographical Information System to make the information accessible. However, due to lack of funding, this has not yet been completed.

In 1996, the School of Architecture undertook further inventory of the architectural and industrial archaeological heritage in the Dublin Docklands Area commissioned by the Dublin Docklands Development Authority. The study was based on the same methods developed for the Built Environment Survey and completed inventories of 61 streets and 188 buildings together with 430 industrial archaeology sites (UCD, 1996). Thus, while the format of the Built Environment Survey forms a basis for further inventories in Dublin, until the information is made accessible it will not be fully utilized.

In 1998, the Historic Heart of Dublin Ltd was established as a partnership with the Dublin Civic Trust and Dublin Corporation under a European Union funded 'Article 10' Urban Pilot Project (via the European Regional Development Fund). The project was mainly initiated as a response to the need for further inventories of Dublin's building stock. Despite the above examples of inventory projects in Dublin, there remains a need for a more uniform approach to the cataloguing of the built heritage.

The integration of urban planning and the historic built environment

The ICOMOS Washington Charter recommended that 'in order to be effective, the conservation of historic towns and other historic urban areas should be an integral part of coherent policies of economic and social development and of urban and regional planning at every level' (ICOMOS, 1987). It further stressed the need for conservation strategies which are based on 'multidisciplinary studies' including archaeology, history, architecture, techniques, sociology and economics. The Amsterdam Declaration identified that 'the legislative and administrative measures required should be strengthened and made more effective' (Council of Europe, 1975) while the Granada Convention reinforced the concept of 'integrated conservation'. Although Ireland signed the convention in 1985, new legislation encompassing a framework for conservation planning and local authority policy on the historic built environment only came into force on 1 January 2000 (MacRory and Kirwan, 2001). The new provisions will have particular relevance for the historic core of Dublin.

The legislative background

The contemporary planning system in Ireland is based on the Local Government Planning and Development Act of 1963, together with subsequent amendments. The Act recommends local authorities to include in their development plans the 'preservation of buildings of artistic, architectural or historical interest' (Nowlan, 1988: 128). In 1976, the Act was revised to include interiors and amended to cover 'preservation of plasterwork, staircases, woodwork or other fixtures or features of artistic, historic or architectural interest and forming part of the interior of structures' (Nowlan, 1988: 128). However, the 1963 Act did not indicate any specific criteria for how conservation should be approached and local authorities have not been obliged to list buildings under this Act. Unlisted buildings, group of buildings, or interiors under the Act may be altered or demolished without any planning permission. However, legislation such as the Housing Act of 1969, which protects buildings in residential use, has provided a certain degree of protection to historic buildings. In addition, the Derelict Sites Act, 1990, has given local authorities powers to prevent buildings from becoming derelict and for levying fines on owners who fail to make required repairs.

Despite these provisions, the fact that many planning departments in Ireland have lacked the resources to provide adequate lists of buildings to be protected has put many historic buildings at risk. In addition, a number of 'exempted developments' were identified in the 1963 Planning and Development Act, with serious negative ramifications for conservation. Developments relating to 'the carrying out of works which affect only the interior of the structure or which do not materially affect the external appearance of the structure so as to render such appearance inconsistent with the character of the structure or of neighbouring structures', were exempted from the requirement to seek planning permission (Nowlan, 1988: 15). Thus, it has been possible to alter the external appearance of a building so long as it is not regarded as a material change and inconsistent with the character of adjoining structures. The replacement of traditional timber windows, with window frames and glazing bars made from Upvc, has been a widespread problem resulting from this provision.

Although regulations have been introduced to take on board restrictions on the exempted developments, the 1963 Act remained vague regarding conservation. Historically, the protection of listed buildings has not been guaranteed because of the lack of both conservation enforcement and penalties for non-compliance. This is particularly relevant within Dublin.

In addition to legislation within the Planning Code there are the National Monuments Acts (1930–94) which empower the Office of Public Works to secure protection of structures, sites or objects of national importance. The

Acts automatically protect any structure pre-dating 1700 and also include archaeological sites. As a result of this enormous task, many post-1700 buildings have been left out of account. Thus, this legislation has been of little significance to most historic buildings in Irish towns (Coleman, 1988). Moreover, an attempt to apply the National Monuments legislation to safeguard a post-1700 historic building by placing it on the Register of Historic Monuments spectacularly failed. The building in question, situated in Capel Street, was one of the few remaining eighteenth-century houses within Dublin with its original internal panelling intact. The building was served with a dangerous structures order despite the fact that estimates for remedial work were less than the cost of demolition and, in February 1993, the façade was removed, leaving the remaining internal architectural features exposed to the elements (Pickard, 1998).

A new legislative programme has been developed jointly by the Department of Arts, Heritage, Gaeltacht and the Islands, together with the Department of the Environment and Local Government. A report in 1996, entitled 'Strengthening the Protection of our Architectural Heritage', forms the basis for this new legislation. It was produced by an inter-departmental working-group with the aim of developing a more thorough form of legislation which would remain within the Planning Code (DACG, 1996) reflecting the Government's commitment to strengthen the capacity of the local government system (DoE, 1998). This has resulted in two pieces of legislation entitled Local Government (Planning and Development) Act, 1999, and the Architectural Heritage (National Inventory) and Historic Monuments (Miscellaneous Provisions) Act, 1999. The former empowers local authorities to enforce conservation by bringing a number of new elements into the legislative framework. First, it makes the listing of historic buildings in local development plans mandatory. Second, once a building has been given the status of protection, the whole building must be protected and any works which might affect the character of the building will require planning permission. Third, it introduces statutory support for 'Architectural Conservation Areas' in order to create a more coherent approach to historic building designation. Fourth, it places greater duties on owners of protected buildings and brings in a new system of penalties against those who neglect them. Finally, it also empowers local authorities to acquire protected buildings or structures if owners fail to maintain them adequately (DoE, 1999c). The latter legislation places the National Inventory of Architectural Heritage on a statutory basis with systematic identification and national standards (DAHGI, 1999). The Heritage Service is currently developing specific guidelines for local authorities in implementing the new conservation legislation.

In addition to the newly developed statutory system, a grant-funding scheme to support conservation work has been launched by the Department of the Environment and Local Government. An annual scheme of grants amounting to approximately £4 million will be distributed to and administered by local authorities. They will also receive special funding to hire conservation expertise into their departments (DoE, 1999a). The grants will be allocated to support owners or occupiers towards conservation of listed buildings of architectural, historical, archaeological, artistic, cultural, scientific, social or technical importance (DoE, 1999b). Thus, years of work towards achieving conservation legislation is finally being reflected in the national government's acceptance, in a limited way at least, of its financial responsibilities for the maintenance and rehabilitation of old buildings.

Conservation policies as reflected in the development plans for Dublin 1971–99

There has been a growing awareness of conservation within the local planning authority in Dublin since the first development plan was adopted in 1971 (Negussie, 1996). Conservation polices embodied in the 1971 Development Plan were poorly developed, reflected in a relatively small number and very limited range of types of buildings which were listed for protection. The plan was prepared and adopted during a time of economic prosperity and widespread acceptance of modern architectural idioms.

In the 1980 Development Plan, the listing system was reorganized and the number of listed buildings increased. The plan also introduced the designation of conservation areas through a land-use zoning policy, but these have meant little as they were introduced without any statutory support. As noted earlier, a report from An Taisce (1985b) showed that by the mid-1980s, a large number of listed buildings had either been demolished or permission had been granted for their demolition.

However, the first major change in attitude towards conservation reflected in the development plans for Dublin can be traced to the 1991 plan. The extension of listings between 1980 and 1991, was more dramatic than previously and the establishment of a separate list for interiors finally demonstrated some acknowledgement of the importance of the extensive range of valuable interiors in the city (Negussie, 1996).

Thus, in Dublin, the protection of old buildings is based on a listing system under the control of the planning authority. In addition to the listing of buildings, the Development Plan also contains separate lists that outline stone setts in streets, granite paving and kerbing, and zones of archaeological interest. However, the listing system has been inefficient for a number of reasons. First, listed buildings have lacked sufficient overall guidance in the development plan, mainly due to an understaffed planning department with little conservation expertise. Second, the system has involved the listing of façades rather than whole buildings and although the listing of interiors has been introduced within the past decade, a relatively low number of interiors have been listed to date. Third, the distinction between different lists of priority resulted in increasingly flexible attitudes towards protection for the majority of the listed buildings (Negussie, 1996; Pickard, 1998). However, such distinction between protected structures is not made under the newly adopted legislation.

The Dublin City Development Plan of 1999 reflects conservation policies that indicate a slightly more integrated approach to the historic built environment than previous plans. This is partly reflected in the increased number of listings. In the current Development Plan, there is also an expression of commitment to the greater use of conservation expertise and the recent appointment of a Conservation Officer marks the first such appointment at an Irish local authority. In addition, there is greater acknowledgement of the importance of co-operation with voluntary conservation bodies in the identification of unlisted buildings. Thus, there is a recognition of the need for a more coherent approach to the historic environment and conservation has become more integrated in the current plan for Dublin than in the past. As a result of the introduction of the Government's annual grant-scheme, Dublin Corporation will receive a figure of nearly IR£1 million to support conservation (DoE, 1999a). Prior to this the only existing local authority support towards conservation in Dublin was a sum of approximately IR£70,000 taken from its own general budget. The integration of planning and urban conservation is moving in the right direction, although there is still a long way to go. However, with the new *Integrated Area Plans* there is now an opportunity for the integration of a wider range of issues, including conservation, at a local level. It remains to be seen whether this approach will be effective in regard to environmental matters.

Urban design policies and the integration of new development

The Dublin City Development Plan, 1999, outlines a general Civic Design Framework with stress on the Liffey quays, a civic thoroughfare, nodal spaces, the integration of large-scale redevelopment areas, pedestrian linkages and radial market streets. Local urban design plans are prepared within the context of the integrated area plans and other local area plans. The general guidelines to these plans are based on 'the development of key urban spaces', 'the use and design of buildings' and 'the development of a strong network of streets'. Concerning the integration of new buildings 'it is the aim of the Corporation to promote a modern architectural expression which is innovative, forward-looking, technically efficient and flexible' (Dublin Corporation, 1999: 77). The Development Plan also states that new buildings and conservation projects must be respectful to the context. This is in accordance with the ICOMOS Washington Charter, in that it identifies that the introduction of contemporary elements in harmony with the surroundings can contribute to the enrichment of an area. In addition, a mixed-use strategy is encouraged in order to achieve a dynamic mixture of uses. With regard to high buildings, the policy is to protect the existing skyline of the inner city, while at the same time recognizing that high buildings may be appropriate in certain areas (Dublin Corporation, 1999). This slightly ambiguous approach to height policy, and the pressure for high-rise developments in the inner city, has led to intensive debate as to what standards should be accepted.

The design guidance provides a framework for what Dublin Corporation intends to achieve within the Development Plan parameters concerning density, height, use and design. These are then implemented through negotiation with developers on their applications for planning permission to develop. In some cases, especially for the redevelopment of the public domain, international design competitions may be held. These are also based on the general design guidelines as stated in the Development Plan. One such competition was held for the development of the urban space at Smithfield. This area forms part of the Historic Area Rejuvenation Project (HARP), which is one of five integrated area plans in Dublin, in the north-west sector of the inner city (McDonnell, 1996). The renewal of Smithfield involves the adaptive re-use of buildings and spaces and the positioning of new traffic routes. The adopted design solution is expected to act as a catalyst for commercial development in an area that contains both historic buildings and derelict sites. In Smithfield Village, for example, old distillery buildings have been adapted as hotels and apartments. However, rather than attempting to imitate the old buildings, new construction incorporates design elements which are adopted from the older buildings, turning them into contemporary features.

Whilst recent developments in Smithfield demonstrate interesting design solutions, they also indicate major changes in the scale and character of the area. The integrated area plan approach aims at strengthening the local character of the area and benefiting the indigenous community. However, the area around Smithfield is currently being exposed to property speculation and there is a sense that the new developments have contributed little to the indigenous community. There is a risk that the speed of development and change may threaten the cultural continuity of the neighbourhood, an area of economic decline and physical decay but also renowned for its fruit and vegetable markets and a horse market held on the first Sunday of every month. It remains to be seen whether the traditional qualities in Smithfield will be lost at the expense of innovation and development.

Management and regeneration action in Dublin

Regeneration of the historic environment in Dublin to date has mostly relied on the private sector rather than on specific conservation-led strategies. While conservation objectives have formed part of Local Area Action Plans, or by the recently introduced Integrated Area Plans, redevelopment rather than renovation has been the priority. As will be discussed in more detail below, the regeneration of Temple Bar has relied more upon cultural and economic development than on the specific conservation of historic buildings. However, there are examples of various small-scale conservation projects in Dublin, mostly resulting from the initiatives of conservation groups. The Green Street Trust is an example of a voluntary effort to restore a 200-year old Debtor's Prison in the north inner city of Dublin. The aim of this project was to contribute to the enhancement of a declined area through the conservation and adaptation of an historic building for working, residential and community uses (Cuffe, 1995).

The local authority has lately recognized the importance of such small-scale projects and is currently undertaking a 'living over the shop' study with the aim of developing a model to demonstrate how conservation could be combined with appropriate uses of historic buildings. Such an approach is deemed necessary to introduce new life and a better mixture of uses in central retail areas. Various surveys have suggested that the proportion of under-used upper floors above retail outlets is high. A study undertaken in 1991 showed that 23 per cent of 198 upper floors surveyed in the George's Street area and South Anne Street, two centrally located commercial streets, were vacant while another 10 per cent were in nominal storage use (Dublin Chamber of Commerce, 1994). Considering the high demand for housing in Dublin's inner city and the rate of new construction to meet these demands, there is a need for the utilization of such under-used floor space.

The case of Temple Bar

The renewal of Temple Bar is an example of cultural-led regeneration facilitated by a market-oriented approach to planning in order to attract private sector investment. This 'entrepreneurial' form of planning has been employed in many other European cities throughout the 1980s and 1990s. It has emerged due to a shift from an emphasis on local authority based planning to central authority planning facilitated by the drawing up of 'general planning frameworks' that promote the private sector (McGuirk, 1994). Furthermore, it has been suggested that this new 'market-sensitive' form of planning changes the relations between planning and market and produces interactions between the two of them instead of oppositions (Healey, 1992). Although conservation policies formed a major part of the initial project scheme for Temple Bar, it has proved more successful in terms of contemporary architecture, the creation of residential space and the development of cultural and tourist activities, often at the expense of the built heritage (Pickard, 1998).

Background of the area

The Temple Bar area is situated in the centre of Dublin on the southern bank of the river Liffey. According to the Roque's map of 1756, many of the street names that exist today can be traced back to this period and much of the medieval pattern remains the same. Most of the old built features originate from the eighteenth century, with Georgian houses and warehouses traced from this time. The area was formerly important for port-related warehouse functions with some streets leading directly to the river (An Taisce, 1985a). It had also been a place for entertainment including theatres and a music hall. During the nineteenth century the area became more commercial and less residential, the port-related functions also moving further downstream. However, the most important changes took place in the 1950s when the area started to lose many of its traditional businesses (Liddy, 1992). The area was subject to further decline and dereliction when the national transport company (CIE) started to buy property in order to secure the site for a planned bus station (Temple Bar Study Group, 1986).

From a conservation viewpoint, the plans for a bus station spelled disaster. Over a period of many years as CIE site assembly took place, important buildings were left to decay as maintenance was terminated. Meanwhile, in order to achieve some rental income from its slowly growing property portfolio, CIE established short-term leases to a wide variety of traders and artists, resulting in a new spontaneous mixture of cultural and semi-commercial activities. In 1986, the Temple Bar Study Group, a voluntary interest group led by students and conservationists, was set up to examine the transportation plans and their possible impact on the environment. It lobbied the planning authority to withdraw its support for the bus station by listing buildings in Temple Bar, refusing to permit the demolition of streets and by promoting the restoration of buildings not in ownership of the national transport company. It was further suggested that an action area plan for Temple Bar should be formulated to provide scope for a more sensitive approach to the protection of the area (Temple Bar Study Group, 1986). In 1991, the plans for the bus station were withdrawn and Dublin Corporation formulated an Area Action Plan for Temple Bar. This was of special importance to groups with conservation interests, especially to An Taisce which had outlined a strategy for Temple Bar through campaigning and undertaking a building inventory of the area.

Government action for renewal

In 1991, the importance of maintaining Temple Bar as a cultural quarter was recognized by the designation of the area as a European Urban Pilot Project following a successful application by Dublin Corporation and a local business initiative called ‘Temple Bar 91’ (Smith and Convery, 1996). This arose in response to the nomination of Dublin as the European ‘City of Culture’ for that year (European Commission, 1994). In addition, two companies were established by the Government under the Temple Bar Area Renewal and Development Act 1991: Temple Bar Renewal Limited and Temple Bar Properties Limited (TBP). The former was set up to vet and approve the payment of financial incentives in order to assure a certain mixture of cultural activities. With the company’s approval an owner would be entitled to fiscal incentives as defined in the Finance Act of 1991. Temple Bar Properties was established as a quasi-state development company acting on behalf of the government for the compulsory acquisition of land in the Temple Bar area. It was to act as a landlord-developer of the area and took over the property owned by the CIE. Its specific aims were the implementation of safety and aesthetics considerations; the renewal and restoration of the streetscape and the building pattern of the area; and the redevelopment of derelict or vacant sites in a way which complemented the conservation of the area (DoE, 1991).

Tax incentives under this scheme provided a 100 per cent tax allowance for refurbishment and a maximum of 50 per cent for construction of new buildings intended for specific approved uses (MacLaran, 1993). The developments carried out by TBP were estimated to cost approximately IR£100 million under the first 5-year implementation period and the private sector was expected to invest about the same amount (DoE, 1994). The budget of TBP was financed by the sale and renting of properties, from European funding and a tax incentive scheme financed by loans from Irish and European banks (TBP, 1992).

The plans for Temple Bar

A Development Programme was produced by TBP as a guiding document to indicate how the area should be developed. The initial statement indicated a strategy to build on what had already taken place spontaneously in the area, recognizing the unique mixture of cultural activities and small businesses which had then come to characterize the area (TBP, 1992). Thus, the aim was to use this as a basis for further development of cultural, residential and business activities over an implementation period of 5 years. The project relied on a marketing-strategy that promoted a certain image of Temple Bar as 'alternative' or 'unique' and as being an area of economic development. As a result of an architectural competition held by TBP, an Architectural Framework Plan was formulated by 'Group 91 Architects'. The plan emphasized the importance of pedestrianization, the creation of public squares and the upgrading of the buildings along the River Liffey quays. Furthermore, it presented a strategy for the conservation of the built environment that suggested 'minimal demolition' and 'imaginative refurbishment'. It also proposed that new buildings should be modern in style and scaled in their context to enrich the existing architecture (TBP, 1991). The Framework Plan had to follow the guidelines set up in the Area Action Plan produced by local authority planners. The physical objectives of this plan aspired 'to conserve the streetscape, layout and building pattern of the area; to retain, through refurbishment, as many of the existing buildings as is practicable; to encourage infill development of a type that complements the scale, height and general appearance of existing buildings; to continue the improvement of pavements, street surfaces and street furniture through a programme of physical conservation' (Dublin Corporation, 1990: 284). Thus, two plans formed the basis of the overall renewal strategy: a flexible architectural framework plan and a statutory local authority action plan.

Aspects of conservation

The approach to the integration of new buildings in Temple Bar has been the use of contemporary architecture at the scale of the historic buildings rather than imitation of the old. The significant feature of Temple Bar is that the historic structure of the area was conserved: the width of the streets and the scale of the buildings were retained with a sharp contrast between 'old' and 'new' in the historic setting (Fig. 8.6).

While the use of contemporary architecture has been successful, the approach to refurbishment has been less convincing. A study by An Taisce identified that twenty distinctive buildings, including a terrace of eighteenth-century houses on Essex Quay, had been totally or substantially demolished by TBP. Furthermore, it stated that several buildings had been destructively treated involving the retention of façades and the removal of important interior features (Smith and Convery, 1996). As with the case of the Children's Cultural Centre, located around the new Meeting House Square in Eustace Street, part of the listed building was demolished in order to make way for a pedestrian passageway. The retention of façades rather than

Fig. 8.6 Temple Bar: the corner of the new Art House and Georgian buildings in Eustace Street.

entire buildings has been identified as a widespread problem throughout the urban renewal schemes in Ireland (KPMG, 1996).

There are various examples of adaptive re-use of historic buildings in Temple Bar, some of which have been more successful than others. The most controversial example is the conversion of St Michael's and St John's Church, the oldest Pre-Emancipation Catholic Church in Dublin. It was converted into a Viking Museum by TBP and Dublin Tourism with the assistance of IR£3 million funding from the European Regional Development Fund (Negussie, 1996). Whilst the removal of the altar, pews and organ was to be expected, wide-scale destruction of significant internal architectural features was not (Pickard, 1998). The action was condemned by An Taisce in a letter to the *Irish Times* by highlighting the fact that all the original plaster and joinery features of the church interior had been removed with the exception of the ceiling and some of the window surrounds (O'Sullivan, 1995). Temple Bar Properties in reply claimed that they were restoring the fabric of an important group of buildings in a manner that would consolidate their use in a sustainable way for the future (Magahy and Magee, 1995). This controversy of appropriate and sensitive re-use of historic churches is not only confined to Temple Bar. Elsewhere in Dublin, there is constant pressure to find new appropriate uses for redundant churches as a result of a decline in the numbers attending services.

The Development Programme for Temple Bar aspired to achieve a socially mixed community of around 2,000 residents. Although gentrification has occurred to a large extent, there is one example of action taken to protect a socially mixed neighbourhood. In 1998, Dublin Corporation decided to purchase fifty-four flats known as the Crampton Buildings in

Fig. 8.7 Temple Bar: Meeting House Square.

order to prevent gentrification of this particular neighbourhood. The current population of Temple Bar is estimated to be 1,200, but with the completion of the final residential schemes in the West End TBP expect to reach their target by the year of 2000 (Coyle, 1999). However, property prices in Dublin have escalated in recent years and it is now hard to imagine a socially mixed community in Temple Bar in the future (Fig. 8.7).

A successful approach to conservation?

There are differing views amongst local planning officials regarding the actions that have taken place within Temple Bar. It may be argued that many new developments do not correspond to the City Development Plan. Certainly it is debatable whether a good balance has been achieved between conservation and renewal. Moreover, conservation groups have expressed the opinion that 'conservation' *per se* has been of low priority. They have argued that too much renewal has taken place at the expense of 'real conservation', with the tax incentives working in conflict with conservation and resulting in the retention of façades rather than the conservation of whole buildings. Furthermore, conservationists have been critical of the fact that the marketing of Temple Bar as a tourist attraction and a cultural centre has actually threatened important buildings in addition to the atmosphere of the area (Negussie, 1996).

These critical views have been confirmed by the government commissioned Study on Urban Renewal Schemes. This concluded that the scheme incentives had significantly contributed to the regeneration of Temple Bar with private sector investment estimated at IR£73.3 million by 1996. It also pointed out that the scheme was unique in relation to other urban renewal schemes in Ireland as the control of tax incentives was restricted to specific uses, thereby guaranteeing some mixture of activities. Furthermore, it showed that 72 per cent of the overall investment was in refurbishment while the average investment in all urban renewal schemes was as low as 11 per cent. At the same time it recognized that proper conservation through refurbishment had not been enforced partly as a result of modern building regulations (KPMG, 1996).

In 1996, An Taisce explained that the conservation record in Temple Bar had been poor due to a number of reasons. Smith and Convery (1996) have cited the following points:

- a lack of specific reference to how conservation should be approached;
- a paucity of conservation expertise;
- a lack of a proper inventory of the area;
- the continued availability of tax allowances for developments replacing demolished buildings;
- the lack of a grants programme to sponsor long-term conservation works; and,
- an absence of any local community partnership.

It may be concluded that the Temple Bar project has proved successful in terms of revitalization, new developments, and economic and 'cultural' (in its widest meaning) regeneration. Nevertheless, urban conservation needs have been under pressure. The standard of conservation work has been compromised in many cases, especially by the introduction of new uses for historic buildings. Thus, a non-restrictive approach to conservation has enabled commercially compatible uses in Temple Bar. The enhancement of older areas usually involves adaptation to modern circumstances and a pertinent question has been raised in this respect: 'should the existing urban forms be adapted or reconstructed to accommodate the new functional demands, or should these demands be constrained to fit into the existing forms?' (Burtenshaw *et al.*, 1991: 140). While the Temple Bar project has resulted in a dynamic mix of uses, it seems that the commercial and tourist aspects of the project have been prioritized at the expense of conservation. The marketing of Temple Bar has to a large extent relied on the tourist sector, which has in turn posed a threat to the conservation of the area. The key question is whether this kind of large-scale profit-oriented urban renewal sits well with conservation objectives. There are various side effects of this type of urban renewal that result in a less sustainable urban development. It remains to be seen whether the introduction of a more integrated area-based approach to planning, as through more recent urban renewal schemes and afforded by the strengthened statutory footing for heritage protection, will resolve the negative effects of large-scale urban renewal.

Environmental management

The greatest environmental threat facing Dublin is its growing problem of traffic congestion. Rapid economic growth over the past decade has contributed to a significant increase in car ownership and use, which has strained the existing road infrastructure. In combination with urban sprawl and an insufficiently developed public transportation system, the city is facing a congestion crisis that demands both short- and long-term solutions. Dublin Corporation has estimated that levels of benzene and nitrogen dioxide pollution may soon exceed European Union recommended top levels (McDonald, 1999b) threatening the overall environment including historic buildings.

Since the 1960s, there has been growing usage of the car as the primary mode of transportation. While levels of car ownership are still low by European standards, it has been suggested that the growing traffic problem is due to a lack of investment in quality public transportation. Consequently, cars are being used in Dublin for journeys that would be covered by public transport in other major European cities (Killen, 1999). Public transport in Dublin is limited to a bus network and an electrified commuter railway line serving the coastal suburbs. Most development plans for Dublin, until very recently, have sought to facilitate traffic movement in the city by road-widening schemes that have often involved the demolition of historic buildings. However, in 1995 the Dublin Transport Initiative was introduced with the aim of creating an efficient public transportation system, including a light railway (LUAS, meaning 'fast' in Irish) and a Quality Bus Corridor Programme. Unfortunately, major delays in the implementation of the strategy and other factors have resulted in a 'transportation deficit' as represented by severe congestion and environmental pollution. A short-term action plan has been introduced to tackle this, including bus, suburban rail, cycling and parking projects (DTO, 1998). Dublin Corporation is also placing new emphasis on pedestrian routes in the inner city, as with the case of O'Connell Street, a main thoroughfare, which is to have the number of traffic lanes reduced in order to widen its footpaths. In addition, a second pedestrian bridge over the river Liffey has recently been built. But there is a deep-rooted problem concerning the separation of land-use and transport planning and a major challenge remains in this respect (McDonald, 1999a).

Tourism and heritage management

Tourism plays a significant role in the economy of Dublin and has become an important element in the redevelopment of many inner-city areas, particularly Temple Bar. This is reflected in the increased number of hotels, restaurants and exhibition centres in the inner city. Some of these developments have seriously affected the historic built environment. For example, despite vigorous conservation lobbying and a third-party appeal to the planning appeals board, developers were granted permission for either demolition or alteration of fourteen historic buildings in order to build a new modern hotel on College Green, facing Trinity College, adjacent to the Temple Bar renewal district. Increased tourist activity is also reflected in the number of historic building conversions supporting the tourist industry, e.g. the conversion of St Andrew's Church into a tourist information centre.

Thus, while tourism has created new opportunities and demands for conservation and adaptive re-use of historic property, it has had some negative impact on the historic built heritage and the character of old areas. There are several examples of conversions of property for tourist use that have seriously altered original structures of historic or architectural significance. On an area level, commercial and tourist-driven urban renewal has also threatened the local character and social sustainability of certain districts.

In the case of Temple Bar, the marketing of the area as a tourist attraction has had a negative impact on the conservation of important buildings and has simultaneously commercialized its former 'bohemian' atmosphere. The conversion of St Michael's and St John's Church into a Viking Museum generated heated debate. Although the conversion guaranteed continued use of the building, the project was not particularly conservation sensitive. Additionally, the character of the area has become significantly more commercialized. A newspaper article suggests that the area has been transformed from a small-scale cultural quarter to a 'shopping and dining Mecca for the affluent tourist' (Coyle, 1999). However, this has to be considered in light of the fact that there are now more visitors in Temple Bar. Furthermore, there is also a larger residential group in the area than before it underwent renewal. The former 'bohemian' atmosphere of Temple Bar was a result of low-rent and short-term leases given out by CIE while transportation plans were in process. Thus, the area was bound to change after the abolition of these plans. Nevertheless, closer co-operation with community and environmental interest groups may have contributed to a more sustainable approach to conservation.

Sustainability

The overall aim of the Irish Government's policy on sustainable development is:

> to ensure that economy and society in Ireland can develop to their full potential within a well protected environment, without compromising the quality of that environment, and with responsibility towards present and future generations and the wider international community (DoE, 1997b).

In 1995, the Department of Environment issued specific guidelines on sustainable development to local authorities in accordance within the aims of the Local Agenda 21 programmes advocated by the Earth Summit (United Nations, 1992). These aims are reflected in the Dublin City Development Plan of 1999, which generally expresses aspirations to reinforce the inner city by incorporating the Government's strategy on sustainable development (Dublin Corporation, 1999).

Amongst the challenging issues facing Dublin are a number of different aspects of sustainability in the historic environment that require immediate attention:

- the rapidity of development, with approximately 4,000 planning applications a year;
- high-rise building proposals in the city centre which may alter the historic skyline of Dublin;
- the continuation of demolition and alteration of historic buildings, including listed buildings;
- appropriate use of the existing building stock, which may demand the relaxation of building and fire regulations for historic buildings;
- an urgent need for solutions to tackle the threat of a growing traffic problem in Dublin.

The present rate of development in Dublin renders it difficult to undertake forward planning. This may result in planning permissions being given to developments that do not comply with overall policies on sustainable development. As regards high-rise developments, although these may increase density in the inner city and are sustainable in terms of better use of land, they do not fit in with the established historic character of the city.

With the adoption of new measures to enforce conservation, future demolition or alteration of historic buildings may be avoided. The new legislation also gives scope for a more area-based approach to conservation that may fulfil the need for further recognition of the historical context of buildings in order to maintain the character of entire areas. As regards adaptive re-use of buildings, the Government strategy on sustainable development stresses the need for re-use of historic property enhancing long-term preservation. In addition, the strategy recognizes the importance of further relaxation of building regulations (DoE, 1997b). Government policy on solutions to the current traffic problem in the inner city promotes closer co-ordination of transport and land-use planning combined with efficient public transport (DoE, 1997b).

Conclusions

Conservation of historic cities is dependent on a sensitive balance between preservation and change in the historic built environment. In Dublin, necessary urban change has been dominated by large-scale renewal. This solution was adopted in response to widespread urban dereliction and decay in the 1980s. In addition, a sudden economic boom in the 1990s alongside a poorly developed framework for conservation has exposed the built heritage to the threats of demolition and alteration. Thus, despite the growing awareness of the built heritage, redevelopment has often been prioritized over conservation. Time will tell whether recently adopted conservation legislation and policy on urban renewal, combined with a general increase in conservation awareness, will ensure a necessary cultural continuity of the historic city of Dublin.

Notes

The author would like to acknowledge views expressed by the following: Dr A. MacLaran (Geography Department, Trinity College Dublin); Mr J. Barrett (City

Architect, Dublin Corporation); Mr P. McDonnell (Chief Planning Officer, Dublin Corporation); Ms S. Roundtree (Conservation Officer, Dublin Corporation); Ms R. MacRory (Department of Arts, Heritage, Gaeltacht and the Islands); Mr P. Pearson and Ms P. Stewart (Dublin Civic Trust); Professor L. Kealy (School of Architecture, UCD); Ms E. Sjöberg (Dúchas); Mr N. Hughes (Local Historian); and Ms J. Helps, (Trinity College Dublin).

References

An Foras Forbartha (1986): *Financing Architectural Conservation*, An Foras Forbartha, Dublin.

An Taisce (1982): *Urbana – Dublin's List 1 Buildings: A Conservation Report for An Taisce, the National Trust for Ireland*, Dublin.

An Taisce (1985a): *Dublin: the Temple Bar Area – A Policy for Its Future*, Dublin City Association of An Taisce, Dublin.

An Taisce (1985b): *Georgian Dublin: Policy for Survival*, Dublin City Association of An Taisce, Dublin.

Bor, W. (1967): 'An environmental policy for Dublin', *Journal of the Town Planning Institute*, 53, 293–6.

Burtenshaw, D. *et al.* (1991): *The European City – A Western Perspective*, David Fulton Publishers, London.

Coleman, I. (1988): *The Preservation of Ireland's Built Heritage: A Preliminary Study*, UCD, Dublin.

Council of Europe (1975): *The Amsterdam Declaration*, Congress on the European Architectural Heritage, 21–25 October 1975.

Coyle, C. (1999): 'Called to the bar', *The Sunday Tribune*, 12 September.

Craig, M. (1975): 'Attitudes in context', in *Architectural Conservation – An Irish Viewpoint*, The Architectural Association of Ireland, Dublin.

Cuffe, C. (1995): 'Dublin – the debtors' prison project', in Burman, P., Pickard, R.D. and Taylor, S. (eds) (1995): *The Economics of Architectural Conservation*, Institute of Advanced Architectural Studies, York, pp. 63–66.

DACG (Department of Arts, Culture and the Gaeltacht) (1996): *Strengthening the Protection of the Architectural Heritage*, Stationery Office, Dublin.

DAHGI (Department of Arts, Heritage, Gaeltacht and the Islands) (1999): *Architectural Heritage (National Inventory) and Historic Monuments (Miscellaneous Provisions) Act*, Dublin.

DoE (Department of the Environment) (1986): *Urban Renewal Financial Incentives*, Stationery Office, Dublin.

DoE (Department of the Environment) (1991): *Temple Bar Area Renewal and Development Act*, Stationery Office, Dublin.

DoE (Department of the Environment) (1994): 'Sub-programme 3: urban and village renewal', in Government of Ireland and Commission of European Communities (eds.) (1994): *Operational Programme 1994/1999 – For Local Urban and Rural Development*, Stationery Office, Dublin.

DoE (Department of the Environment) (1997a): *1998 Urban Renewal Scheme – Guidelines*, Stationery Office, Dublin.

DoE (Department of the Environment) (1997b): *Sustainable Development: A Strategy for Ireland*, Stationery Office, Dublin.

DoE (Department of the Environment) (1998): *Statement of Strategy*, Stationery Office, Dublin.

DoE (Department of the Environment) (1999a): *Environmental Bulletin*, issue 43, Stationery Office, Dublin.

DoE (Department of the Environment) (1999b): *Explanatory Memorandum*, Stationery Office, Dublin.
DoE (Department of the Environment) (1999c): *Local Government (Planning and Development) Act*, Stationery Office, Dublin.
Dublin Chamber of Commerce (1994): *New Life for Old – Urban Renewal in Dublin*, Dublin.
Dublin Corporation Planning Department (1986): *Inner City Development – New Incentives for Designated Areas*, Dublin Corporation, Dublin.
Dublin Corporation Planning Department (1990): *The Temple Bar Area – Action Plan 1990*, Dublin Corporation, Dublin.
Dublin Corporation Planning Department (1999): *Dublin City Development Plan*, Dublin Corporation, Dublin.
DTO (Dublin Transportation Office) (1998): *Transportation Review and Short-term Action Plan*, Dublin Transportation Office, Dublin.
European Commission (1994): *Urban Pilot Projects*, Office for Official Publications of the European Communities, Luxembourg.
Healey, P. (1992): 'The reorganisation of state and market in planning', *Urban Studies*, 29 (3/4): 411–34.
Heritage Council (1998) *The Role of the Heritage Council in the Planning Process: Pre-publication Draft*, Dublin.
Horner, A. (1985): 'The Dublin Region, 1980–1982: an overview of its development and planning', in Bannon, M.J. (ed.) (1985): *The Emergence of Irish Planning 1880–1920,* Turoe Press, Dublin.
ICOMOS (International Council on Monuments and Sites) (1987): *The Washington Charter*.
Kealy, L. and O'Rourke, M. (1995): 'Historical endowment: the use of inventories', in Convery, F. and Feehan, J. (eds) (1995): *Assessing Sustainability in Ireland*, Environmental Institute, UCD, Dublin.
Kearns, K.C. (1983): *Georgian Dublin: Ireland's Imperilled Architectural Heritage*, David and Charles, Newton Abbott.
Killen, J. E. (1999): 'Transport: a major problem and issue', in Killen, J.E. and MacLaran, A. (eds) *Dublin: Contemporary Trends and Issues for the Twenty-first Century*, Geographical Society of Ireland Special Publication 11, Dublin, pp. 93–106.
Kirwan, S. (1998): 'Legislation on protection of the archaeological heritage in the republic of Ireland', in Deevy, M. (ed.) *The Irish Heritage and Environment Directory: 1999*, Archaeology Ireland, Bray.
KPMG (1996): *Study on the Urban Renewal Schemes*, Stationery Office, Dublin.
Liddy, P. (1992): *Temple Bar – Dublin*, Temple Bar Properties Ltd., Dublin.
McDonald, F. (1989): *Saving the City – How to Halt the Destruction of Dublin*, Tomar Publishing, Dublin.
McDonald, F. (1999a): 'Road pricing looks likely route in the future', *The Irish Times*, 31 August.
McDonald, F. (1999b): 'Cars in future likely to run on pollutant-free fuel system', *The Irish Times*, 1 September.
McDonnell, P.F. (1996): *Historic Area Rejuvenation Project*, Dublin Corporation, Dublin.
McGuirk, P. (1994): 'Economic restructuring and the realignment of the urban planning system: the case of Dublin', *Urban Studies*, 31 (2): 287–308.
MacLaran, A. (1993): *Dublin: The Shaping of a Capital*, Belhaven Press, London and Newhaven.

MacLaran, A. (1996): 'Office development in Dublin and the tax incentive areas', *Irish Geography*, 29 (1): 49–54.

MacRory, R. and Kirwan, S. (2001): 'Ireland', in Pickard, R. D. (ed.) (2001): *Policy and Law in Heritage Conservation,* Conservation of the European Built Heritage Series, Spon Press, London.

Magahy, L. and Magee, F. (1995): 'The Temple Bar conversion', *The Irish Times*, 6 January.

Mawhinney, K. (1989): 'Environmental conservation concern and action, 1920–70' in Bannon, M.J. (ed.) *Planning: The Irish Experience*, Wolfhound Press, Dublin, pp. 86–102.

Negussie, E. (1996): 'Changing attitudes towards urban conservation in Ireland', Master's thesis, Department of Human Geography, Stockholm University.

Newman, P. and Thornley, A. (1996): *Urban Planning in Europe*, Routledge, London and New York.

Nowlan, K. B. (1982): 'Problems and achievements – south', in *Ireland's Architecture – A Shared Heritage*, conference summary.

Nowlan, K.I. (1988): *A Guide to Planning Legislation in the Republic of Ireland*, Leinster Leader Ltd., Naas.

O'Sullivan, J. (1995): 'Temple Bar conversion', *The Irish Times*, 11 January.

Pickard, R.D. (1998): *Meeting the Requirements of the Granada Convention; A Review of Policy for the Protection of the Architectural Heritage in the Republic of Ireland*, RICS Research Paper Series, vol. 3, no. 1, Royal Institution of Chartered Surveyors, London.

Rowan, A. (1980): 'The historic city', in Nowlan, K., Robinson, N. and Rowan, A. *Dublin's Future: the European Challenge: A Conservation Report for An Taisce*, Country Life, London, pp. 2–7.

Smith, M. and Convery, F. J. (1996): *Ireland's Designated Areas – Lessons from Temple Bar*, An Taisce discussion paper, Dublin.

TBP (Temple Bar Properties Ltd.) (1991): *Temple Bar Lives – Winning Architectural Framework Plan*, Temple Bar Properties Ltd., Dublin.

TBP (Temple Bar Properties Ltd.) (1992): *Development Programme for Temple Bar*, Temple Bar Properties Ltd., Dublin.

Temple Bar Study Group (1986): *Temple Bar Study – A Reappraisal of the Area and the Proposed Central Bus Station*, Dublin.

UCD (School of Architecture) (1996): *Inventory of the Architectural and Industrial Archaeological Heritage – Vol. 1*, School of Architecture, University College Dublin.

United Nations (1992): *Conference on Environment and Development (the Earth Summit)*, Rio De Janeiro.

Giorgio Gianighian

Venice, Italy

Introduction

The *centro storico* of Venice can be seen as emblematic of all the Italian historical cities. The reasons for this choice are clear: there is a concentration of monuments of great beauty in Venice, and its unique characteristics make it difficult to maintain and manage. Nevertheless, in spite of this uniqueness, the principles, which guide conservation in Venice, apply also to other historical centres as well.

Historical background

A rapid examination of the history of the urban development of Venice and of its architecture will help identify some features which form the basis for the city's conservation and at the same time identify the most serious problems that remain to be solved. A first hypothesis on the urban development of the *Serenissima* was that put forward by Trincanato (1971). Crouzet Pavan (1992) develops a fairly exhaustive analysis of the lines of growth of the city on the eve of the Renaissance. For the following period, examples of how the later urban reclamations were executed are found in Pavanini (1989).

One crucial factor in the evolution of the city has always been water, 'acqua dolce', drinking water. 'Venice is in the water, and has no water' is the famous and concise paradox coined by the Venetian chronicler Marin Sanudo, at the beginning of the sixteenth century. To that we could add, less elegantly but not less accurately, that Venice, composed as it is from a multitude of islands scattered 'over the salt waters', also has very little land.

The city's location is so difficult that almost everything was lacking, even drinking water, which had to be collected and conserved through an ingenious system of filtration tanks. It is not by chance that Fernand Braudel is one of the few historians who have taken this fact into consideration: as early as 1979 he published a drawing showing the design of this system (Braudel, 1979: 195).

There were once about 7,000 public and private cisterns in Venice. The cisterns were constructed by excavating a large hole of varying dimensions, but of a depth not exceeding 5 metres; the hole was rendered impermeable by lining with clay, and it was then filled with sand taken from river-beds. A well-shaft of brick was constructed in the centre of this tank, brought up to street level, and capped by the stone wellhead. The pavement, often made of *masegni* (to which we will return later), was the catchment area, and was sloped in such a way that rainwater flowed to the collecting outlets, which were linked to a brick-vaulted drain; this provided the first stage of filtration. The second stage consisted of the river sand; the water filtered through it by gravity and collected around the bottom of the well-shaft. In times of drought, the cistern was used as a holding-tank for water brought to the city from rivers on the mainland.

As the city's population increased – for example, around the middle of the sixteenth century Venice had more than 150,000 inhabitants, possibly reaching a maximum of 175,000 (Beltrami, 1954) – the lack of living space was counterbalanced by the necessity to retain ample open spaces: *cortili* inside palaces, squares and public courtyards, shared courts in the working-class developments. These spaces were necessary to accommodate, beneath the pavement, one or more cisterns. It was only in the sixteenth century that the solution to the problem of fresh water supply was found which did not require the use of undeveloped land. Water was obtained through a system of gutters and pipes which conveyed rain from the roof into a cistern located inside the building (Gianighian, 1989).

We must also note that Venice had to dispose of a notable quantity of various kinds of detritus. The city lives and grows in its lagoons, defined by a fine network of canals, some deep, some shallow, which are vital for communication and commerce. The canals and lagoons were surveyed and monitored with minute accuracy and subjected to daily maintenance in order to keep them clear of detritus and rubbish and to conserve their depth. A continuous programme of excavation of the canals and certain parts of the lagoon accomplished this supervision, and this produced mud and other refuse needing disposal. On the other hand, every building site, especially if it was a matter of reconstruction, had a need for *sacche* (enclosed areas of water) in which waste building materials (*rovinazzi*) could be deposited. These materials were sometimes taken to sea but more frequently found a more controlled and interesting destination in the reclamation of mud-banks (*barene*) (Fig. 9.1), which were being transformed into new urban land, the only possible means of expansion in this bizarre city enveloped by its waters (Pavanini, 1989). It is necessary to keep in mind this aspect of the excavation of canals since it is the cause of one of the problems faced by recent administrations of the city.

A third aspect, once again connected to the site on which the city rises, is that of the conservation of building materials: each element of a structure was conserved for as long as possible, and, in cases of reconstruction, re-

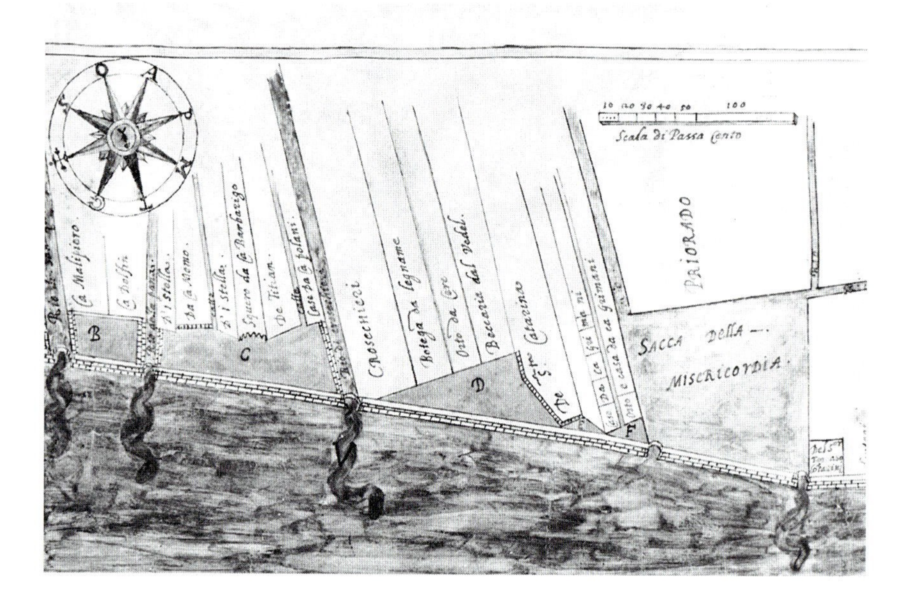

Fig. 9.1 Detail from the original design of land reclamation of Fondamente Nuove, by Gerolamo Righetti, Venice 1590 (private collection).

used. In a palazzo, for example, walls and floors were frequently retained and only the façade was modernized. When Marin Contarini built a new house for himself, one of the most sumptuous in the city (and the most famous: the Ca' d'Oro), he painstakingly re-used on the site every brick and stone from the earlier palace. If a building had burnt down as many elements as possible were salvaged for the new construction.

The reasoning behind this economy is clear: building in Venice poses major construction problems that do not exist on the mainland; building foundations alone, for example, is a complex and costly operation. Furthermore, all of the materials came from far away and had to be laboriously transported by water. Thus we see how it became usual in the city to practise the most rigorous conservation of all elements of a building. In this respect methods of construction and the organization of construction projects have been explored by a number of commentators (Wirobisz, 1965; Connell, 1988; Gianighian, 1990).

The reasons why Venice has survived until today with an almost unaltered urban texture are complex and in part imponderable, but the ancient custom of the saving of building materials, rooted in the past, is certainly an essential element. This remained the practice until the fall of the Republic (1797) and beyond, although for different reasons. No longer a capital city, no longer the *Serenissima*, Venice became impoverished as part of the Hapsburg dominions. Its economy declined rapidly, partly because of the Austrian government policy to promote the development of the port of Trieste. As is so often the case, poverty helped architectural and urban conservation. With the exception of a few grandiose French schemes, such as the *Ala Napoleonica* in Piazza San Marco, the gardens in Castello, the opening of Via Eugenia (the present Via Garibaldi), the two occupying powers initiated few urban projects.

It was after it became a part of the Kingdom of Italy in 1866 that Venice experienced a new economic development. The port was revitalized to rival Austrian Trieste, while business investment, both Italian and foreign, resulted in an increase in the number of industrial establishments, in different sectors such as shipyards, glassworks, textile factories, with the manufacture of clocks, of matches, and so on. As it became wealthier, the city began to modernize, widening some of its streets and constructing important public buildings, including cotton factories, the stock exchange, public slaughterhouses and gasworks; the *Arsenale* was enlarged, hotel provision increased, all at the cost of the loss of the ancient urban texture. From the end of the last century, large numbers of apartment blocks were developed for the working classes, at the instigation of various bodies, both public and semi-public (Romanelli, 1988).

In the meantime, the cultural debate about conservation, always particularly lively in Italy, led to different approaches, often not entirely acceptable, but certainly not to be discounted. Venice saw the transformation of several palaces, by then in an advanced stage of deterioration, into neo-Gothic or neo-Byzantine style, according to a concept of restoration based on the theories of Viollet-le-Duc. The Fondaco dei Turchi well represents all of these: it was subjected to a restoration so radical as to make it unrecognizable if compared to images of the building before the works began. The interventions of Baron Franchetti at the Ca' d'Oro also gave rise to considerable dissent, although today they appear less destructive than our first example (Pertot, 1988). Great debates also arose concerning the restoration work on the Basilica of San Marco, which the disciples of Ruskin claimed was producing something which was purely, and simply, a fake.

Naturally, the process of modernization is not unique to Venice; in fact, what took place here was far less devastating than the storm of change which seized and transformed other Italian cities, such as Turin, Florence and Rome. These cities were subjected to special pressure because of their status as successive capitals of the Kingdom of Italy, and in consequence they experienced more extensive transformation and lost larger parts of their historic urban fabric. By contrast, the work carried out in Venice meant the demolition of only a limited number of existing buildings: for the construction of new buildings, whether for housing or industry, preference was given to extensive areas of undeveloped land or land occupied by earlier, now obsolete, industrial or crafts-based activities. The most significant example of this is the redevelopment of the numerous and extensive *chiovere*, land at one time used for spreading fabrics out to dry after dyeing. It can be stated, therefore, that the city at the end of World War II was mostly unchanged. Even if its built-up areas had been enlarged, it was unaltered in its essential architectural characteristics and, above all, the unique urban structure that makes the city an unrepeatable phenomenon in European history was still conserved. It must be emphasized, however, that since national unification there have been numerous interventions on individual buildings, which, although they have not changed the complex general configuration of the city, have frequently irreparably compromised individual structures of great historic importance.

The policy and planning framework

Conservation between theory and practice

Meanwhile, the debate concerning the future of the historic city centres became steadily more interesting. It is noteworthy that in this respect Italy was in the European vanguard. As early as 1931 the *Carta Italiana del Restauro* (Italian restoration map) was drawn, and then, in 1939, two laws were passed, both in many ways advanced for the time, to protect the national heritage, both architectural and landscape. One must remember here Giovannoni, both as a historian of urban restoration and as a teacher (see Choay, 1988).

After World War II, in-depth studies into the morphology of selected Italian cities were carried out, which contributed towards a climate of awareness and respect towards these historic centres and the buildings of which they were composed. This new awareness found formal expression in the Congress of Gubbio, organized by ANCSA (National Association for Historical and Artistic Centres) in 1960. Subsequently, the debate was further developed, with many new authors taking part; among them Benevolo in Rome, Michelucci in Florence, Pane and Di Stefano in Naples. However, the political classes showed little interest in ANCSA's new approach, and the professional world was equally uninterested. The debate was kept at a high level, as were the laws, which, however, were not enforced. The failure to apply the planning laws throughout Italy left the way open for a wave of speculative building which eroded the historic centre. Venice was not spared, and many historic buildings were demolished and reconstructed.

The city's Master Plan (*Piano Regolatore Generale*) (PRG) was approved in 1962. For the historic centre it provided only for the re-opening of certain canals and *calli* (streets or lanes) and the building of a few new bridges. It postponed the drafting of criteria for implementation to the next step, that of the preparation of the Local Development Plans (*Piani Particolareggiati*) (PP), as provided for in the planning law of 1942. These were finally adopted in December 1975; until that date it was possible to demolish and rebuild without much difficulty. Controls were made the responsibility of the city council's Building Commission (*Commissione edilizia comunale*), headed by an engineer, and including members representing the professional associations.

However, the Building Commission's awareness of the historic heritage was meagre and generally based on little cultural knowledge. Thus it was possible, at the end of the 1940s, to allow the destruction of the group of small houses which stood on the basin of San Marco next to the Prisons, in order to provide space for the new Danieli hotel. This damaged the character of the prospect of the *Riva* (quay) in a particularly sensitive location. Commenting on this episode, Lorenzetti (1974: 284) stated: 'The row of low houses . . . was a reminder of the assassination of Doge Vitale Michiel I (1102), since the murderer found refuge on the spot. The *Signoria* then ordered the houses to be pulled down and banned the future erection of taller buildings'. In this way, too, the Hotel Bauer was able to extend its earlier neo-Gothic palace (facing the Grand Canal) with a new addition on Campo San Moisé, thus completely destroying the character of that square. The fine Palazzetto Foscari on the Grand Canal was demolished, while at the same time strong opposition was voiced, in the name of conservation, to Frank Lloyd Wright's proposed building on the Grand Canal, and to Le Corbusier's project for the new hospital. The few private buildings that were at the time listed and subject to the guardianship of the *Soprintendenza ai Beni Architettonici e Ambientali* (the National Board of Control) could still be 'emptied' behind their façade and rebuilt, since only the exteriors were protected. Nor should we forget the impact of modern technologies on historic structures, in particular the use of reinforced concrete which, if not carefully used, disturbs the delicate static equilibrium of traditional Venetian construction. These few examples should be sufficient to demonstrate the complete failure of the city to carry out choices appropriate to its history.

The plan to restore the historic centre of Bologna (produced from the PRG of 1969), which was born into this contradictory universe, was characterized on one hand by increasingly sophisticated theoretical concepts of restoration: by the study of historic cities, by the actions of ANCSA, by the first tentative steps towards urban restoration. But on the other, by the free hand given to speculative development, which put at risk the very survival of these same historic centres. This plan, rooted in the theories of Leonardo Benevolo, is so celebrated that it requires no further discussion here. It is sufficient to say that it brought in the idea that historic dwellings could be easily adapted to present-day residential needs, when returned to their original form and equipped with modern services. It marked the birth of so-called 'typological restoration'. The principal typologies that characterized the buildings of the city were analysed and classified, and then a programme of building restoration applied to each building. The process included the removal of later accretions and the reconstruction of lost elements, according to what were presumed to have been their original form.

Even if the idea of re-using traditional terrace housing remains interesting and valid, some of the methods used are more debatable, i.e. the removal of parts added in successive periods, stamped with the infamous mark of inauthenticity; the sometimes radical substitution of ancient building materials; and the possibility of rebuilding elements lost or never completed. The plan considers each building as a part, a fragment, of the great monument that is the city itself; no element is either historically or formally autonomous, but is connected to that whole which is the 'city-monument'. This allowed for the unjustified substitution of modern for historic materials, or rather, for reconstruction on the basis of abstract typological 'schemes' and on the basis of theories of restoration that were certainly open to debate, and which sometimes resulted in stylistic falsehoods.

The lack of recognition of the differences between the various compositional and historic elements on the one hand and, on the other, enforced homogenization on the basis of these typological abstractions, constitute the principal limits of a plan. This has, however, considerable merits as far as practical planning is concerned, especially in its emphasis on the relationship between the historic centre and the periphery, the removal of damaging activities, the reorganization of city traffic and so on.

A 30-year-long development plan

This digression was necessary in order to better understand the *Piani Particolareggiati* of the historic centre of Venice, drafted on the lines of the experience in Bologna. But before describing the plans, it is worth drawing attention to an appreciation of the demographic context in which they are set. The population of the historic centre of Venice (including the Giudecca) has fluctuated in the following manner:

1881:	133,000	
1951:	176,000	(including refugees from the mainland)
1961:	137,000	
1971:	108,000	
1981:	92,000	
1992:	75,870	

The most recent data indicate further losses: by June 1999 the population had declined to 67,326 inhabitants.

In these figures we see an emergency, a population exodus resulting from many factors, not least the fears aroused by the disastrous floods of November 1966. Venice became nationally important, so that in 1973 a new Special Law for Venice was promulgated, which put at the disposal of the city extraordinary special funding, and required the drafting of the *Piani Particolareggiati* already envisaged in the PRG of 1962. These local plans had to develop for each part of the city's territory what had been forecasted by the earlier PRG. The PRG is structured into a complex set of executive instruments: the PP; the division of the city into zones of different historic interest; the indication of rigidly defined future categories of use; the definition of the types of intervention; the excessively strict respect for the building unit, established abstractly on the basis of summary topographical analysis; and going as far as the *Piani di Comparto*, in which the authority was required to predict all possible interventions as well as the manner of their execution. It is configured, therefore, like a complex series of Chinese boxes, the last of which will always remain unattainable. The administration therefore found itself confronted by a mass of work of monstrous proportions. To give just one example of this impossible position, of the 405 *Piani di Coordinamento*, the directives of which had to precede the drafting of the *Piani di Comparto*, barely a dozen were actually drafted.

Thus extreme complexity led for obvious reasons to a check mate: in the absence of adequate planning instruments, which awaited the development of the *Piani di Coordinamento*, almost nothing could be done beyond the modernization of sanitary provision, installation of central heating, and routine maintenance. But, in the meantime, the city continues in any case to change, as a result of initiatives (or rather, non-intervention) from public bodies, or from private organizations. In its enforcement of the law, the Buildings Commission behaved in a fairly permissive manner, allowing numerous abuses of the regulations, which were later legalized through amnesties in 1985 and in 1994. These factors explain the increased number of residential units, which in the last 10 years rose from 36,000 to 40,000, in spite of a falling population.

In December 1992, the city council, realizing the impossibility of drafting detailed plans, authorized a new variant (Fig. 9.2) of the 1962 *Piano Regolatore Generale*. This was approved by the City Council of Venice in December 1992 and superseded all previous legislation. This planning instrument, still highly questionable, provided for the classification of all the buildings in the city on the basis of typological references, in which the criteria for approving a proposed intervention were based on an historical-typological form (*scheda*). The classification of buildings follows chronological criteria (pre-nineteenth-century buildings, nineteenth- and twentieth-century buildings), and analyses the (presumed) degree of conservation of their original character. On this basis the degree of strength of legal protection is established: the older the structure, the stronger the protection. It is this theoretical basis that is one of the weakest parts of the entire system: the *schede* are in fact very frequently imprecise or even incorrect in the way they define the building complexes. This is, after all, understandable in a study that claims to cover the entire city.

The reference back to a presumed original situation against which each proposed intervention must be measured, relating the work in question to this original state, serves only to complicate matters further. In a city like Venice, developed in successive layers, built on top of one another, with a notable persistence of styles and of construction traditions, even the simple dating of the earliest phase of construction of a building is almost invariably a task which is both complex and uncertain. The most recent conservation theories, as has been noted, are far removed from any concept of returning to a presumed original 'purity', confirming strongly the necessity to conserve that which the past (in all its guises) has bequeathed to us. The rich patrimony of our historic inheritance, the result of centuries of the life that has animated our cities, is something that does not belong to us. We are simply the custodians, and for which we must account to the whole world, not just today, but above all, tomorrow.

Another aspect of the plan has rendered the need for further alterations: it completely ignores the historic buildings of the nineteenth and twentieth centuries (Fig. 9.3), and concludes with the paradoxical condemnation to demolition of the entire stock of social housing built from the end of the nineteenth century to 1960. A total of at least 2,500 apartments would therefore have to disappear (in accordance with which juridical principle?), because they differ from a typological-morphological point of view from the context of the Venetian built environment, to be rebuilt (with funds from where?) following building types more respectful of local tradition. If

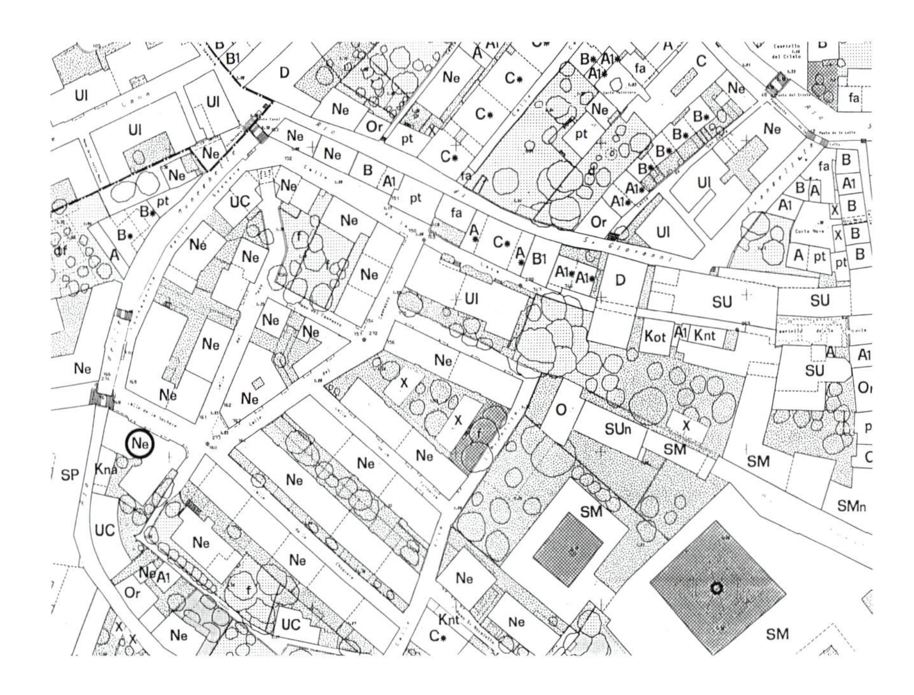

Fig. 9.2 *Variant to the Venice masterplan (1992): detail of the Frari area, where every building is classified according to age (highlighted: the Domus Civica–1921).*

Fig. 9.3 *Domus Civica, which, because of lesser architectural value, could be totally emptied and rebuilt, saving the fronts: a clear example of façadism.*

one was seeking to justify this planning monstrosity one might suppose that we are talking about buildings in poor or unstable condition, but the converse is true; these are sound, well-built dwellings, with materials, including Venetian terrazzo, which nowadays can no longer be used because of their excessive cost. No, the only motive for their condemnation is aesthetic; they are ugly buildings! Further comment is superfluous. We know well how questions of taste can be fickle, variable, subject to cultural fashions and to personal and historical prejudice. They should certainly not be allowed to constitute fundamental values on which to base a discussion on conservation, which is to have the force of law.

The concept of *ripristino* and of typological restoration is based on an anti-historical perspective: why give special privileges to the earliest periods, obliterating the impact made by succeeding generations on a building? It also poses serious conservation problems, especially with Venetian residential complexes, which are organized in a particularly ingenious co-ordinated manner, and have been subjected to such alteration over the past centuries that any attempt to obliterate them today would result in significant loss of historic material.

This urban development plan, in its turn, has been superseded by a new PRG, which has amended, with Leonardo Benevolo's assistance, the more striking errors of the previous version. The general principles of the plan, however, insofar as it relates to conservation intervention, remain unchanged.

Management and regeneration action

Works of restoration: finance

There are, however, other points, which must be emphasized. Venice, like other historic Italian cities, is subject to a law that makes the city's administration responsible for confronting the most serious problems facing the city. This Special Law (no. 798) dates from 1984, but it has been updated on a number of occasions.

First, an illustration of how the funds assigned to the city under the special legislation for Venice have been administered: Lit. 1,654 millions (approx. £570m; US $923m) (Comune di Venezia, 1998). Of this sum, 94 per cent has been allocated, 69 per cent has been committed, and 37 per cent actually disbursed; 5 per cent of the projects are under contract, 54 per cent are currently in progress, and 41 per cent of the works have now been completed.

The projects that are being financed by a complex series of four laws over the period 1992–97 are as follows:

A. Integrated works to canals

These are provided for:

- the excavation of *rii* (the minor canals) and the removal of mud and other deposits;
- the organization and restoration of the existing sewer network;
- the static stabilization of public and private buildings along these same canals; and,

- the protection of urban areas subject to *acqua alta* (flooding by abnormally high tides) by means of raising pavement levels.

The overall project covering both the historic centre and the islands provides for an expenditure of Lit. 1,500 million (approx. £517m; US $837m) over 23 years, from 1994. The *Comune* has assigned responsibility for this work to a company in which it holds majority interest. The first site for these works is the islet of Santa Maria Zobenigo, which includes the Fenice theatre, which was almost completely destroyed by a terrible fire on 29 January 1996. For this phase of the project work in progress comprises 62 per cent of the total, that completed 36 per cent.

The eleven interventions that have been approved, eight of which are in the historic centre, cover canals with a total length of 15.853 km, embankment walls with a total length of 12.632 km. in the public sector and 20.189 km in the private, and a volume of material to be excavated of 102,172 cubic metres. The work on paving (of streets, quays and squares) covers 53,813 square metres, and the bridges affected total 156.

The total length of canals in the historic centre, together with Murano and Burano, amounts to 49.279 km, the length of public embankment walls to 33.290 km, and private 62.830 km, with a final total of materials to be excavated of 338,000 cubic metres. The total extent of paving works is around 53,813 square metres, and there are 454 bridges in total. The average cost of restoring the embankment walls is Lit. 1,039,089 (approx. £358; US $580) per linear metre, according to data provided by the agency responsible for these works, *Insula SpA*.

B. Residential works (already financed by Law no. 798/1984)

These works are designed to arrest and reverse the population exodus, to which are also added cultural projects, including the Venetian museums system and, more recently, the reconstruction of the Fenice theatre. In the residential sector, the objective is to bring back into use degraded properties, including the recovery of *Comune*-owned property at present under-utilized. A further object is the recovery and redevelopment of abandoned industrial areas. From 1993 to the end of 1998 financial support provided for work on a total of approximately 1,000 dwellings, including those for old people and students. To be added to this work, which covers, as can be seen, both restoration and new construction, is the acquisition of apartments and the maintenance of accommodation owned by the *Comune*, amounting to 749 units, with a total area of 52,000 square metres. Of the total fund allocated of Lit. 201.799 million (approx. £69.6m, US $113m) 84 per cent has been committed, and 60 per cent spent, with 57 per cent of work in progress and 43 per cent completed.

A number of other specific actions under this heading are supported:

- For the museum heritage of the city: financial support is being distributed over eleven buildings; the planned allocation is for Lit. 95.353 millions (approx. £33m, US dollars 53m), of which 65 per cent is committed and 47 per cent expended. Work in progress stands at 50 per cent and that completed at 44 per cent.

- Theatres, cinemas and exhibition spaces: this includes a total of nine projects, and an allocation of Lit. 168.996 million (of which Lit. 106m. is for works to the Fenice theatre), and from which to date commitments of 89 per cent have been made, but only 10 per cent spent. The state of progress of this work is unclear, because of the legal appeals concerning the reconstruction of the theatre, which mean that only recently has work there been re-started.
- Schools: this concerns maintenance and restoration as well as new construction work, both in the city and in the islands. In this group the finance is Lit. 49.498 million (approx. £17m, US $27m), of which 67.5 per cent is committed and 43 per cent spent. Work in progress stands at 28 per cent; works already completed at 63 per cent.
- Buildings of public interest: this refers to the more important *palazzi* and monumental buildings (Figs 9.4 and 9.5) that are the responsibility of the *Comune*, with a relevant allocation of Lit. 272.921 million (approx. £94m, US $152m), of which 62.5 per cent has been committed, but only 13.5 per cent expended. Unfortunately, the data relating to the state of progress of the work and the relative percentages have been reported incorrectly in the *Comune's* document, so that it is not possible to establish a clear picture.
- Works of urbanization: this refers to interventions in respect of street paving, parks and other public open spaces, street furniture, the re-ordering of Piazzale Roma, sewage plants, public street lighting and various other works which are the responsibility of the *Comune* of Venice. Within this grouping, finance of Lit. 170.476 million (approx. £59m, US $96m) has been allocated, of which 69 per cent is committed and 49 per cent spent. Work in progress comes to 57 per cent and that completed 26 per cent.
- Contributions towards the private housing stock: this important heading concerns both the restoration of housing and former factories, as well as aid for the acquisition of 'first homes', and for the recovery of ground floor properties subject to flooding. This section has received an allocation of Lit. 200.952 million (approx. £69m, US $112m), of which 90 per cent is committed and 28 per cent spent. Work in progress stands at 18 per cent and that completed at as much as 80 per cent.

The final section, covering *bases of industrial production*, has little relevance to the historic centre of Venice, as does that relating to *sports facilities*.

These figures all indicate very clearly that the public administration has, finally, begun to move, with many projects dedicated to the safeguarding of the city and its revitalization, after decades of paralysis and lack of initiative. On the other hand, the very size and scope of the enterprise means a risk for the health of the built patrimony: damage could result from insufficiently detailed control and care over the quality of the execution of the works.

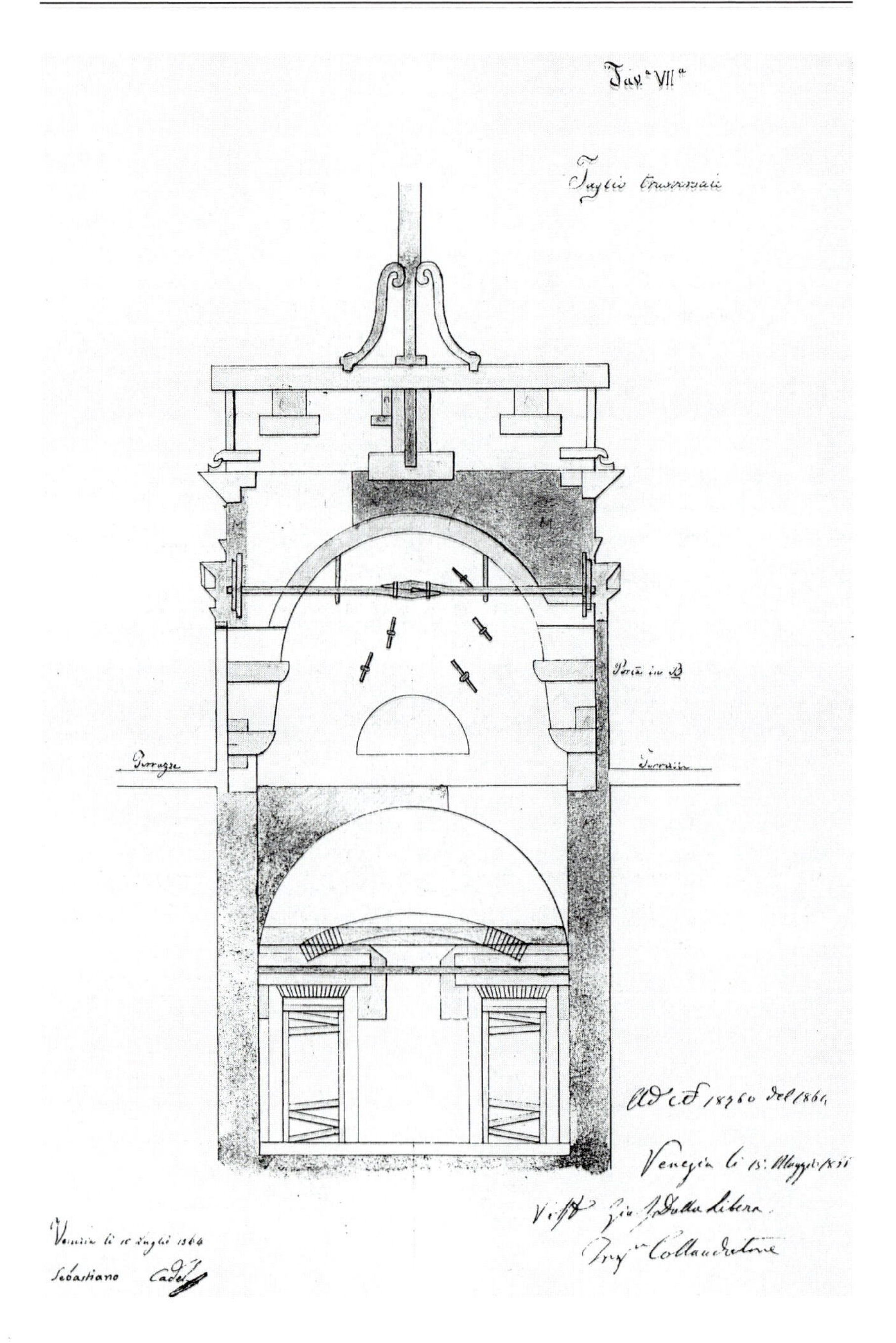

Fig. 9.4 The restoration of the Clock Tower in St Mark's square, 1858: cross section of the upper part.

Works of restoration: the interventions

The physical fabric

Altogether, therefore, the management of this funding has brought notable advantages, above all to the architecture that at one time was considered 'minor'. There have been, however, certain negative effects, in particular the systematic destruction of historic plasterwork, the loss of parts of ground

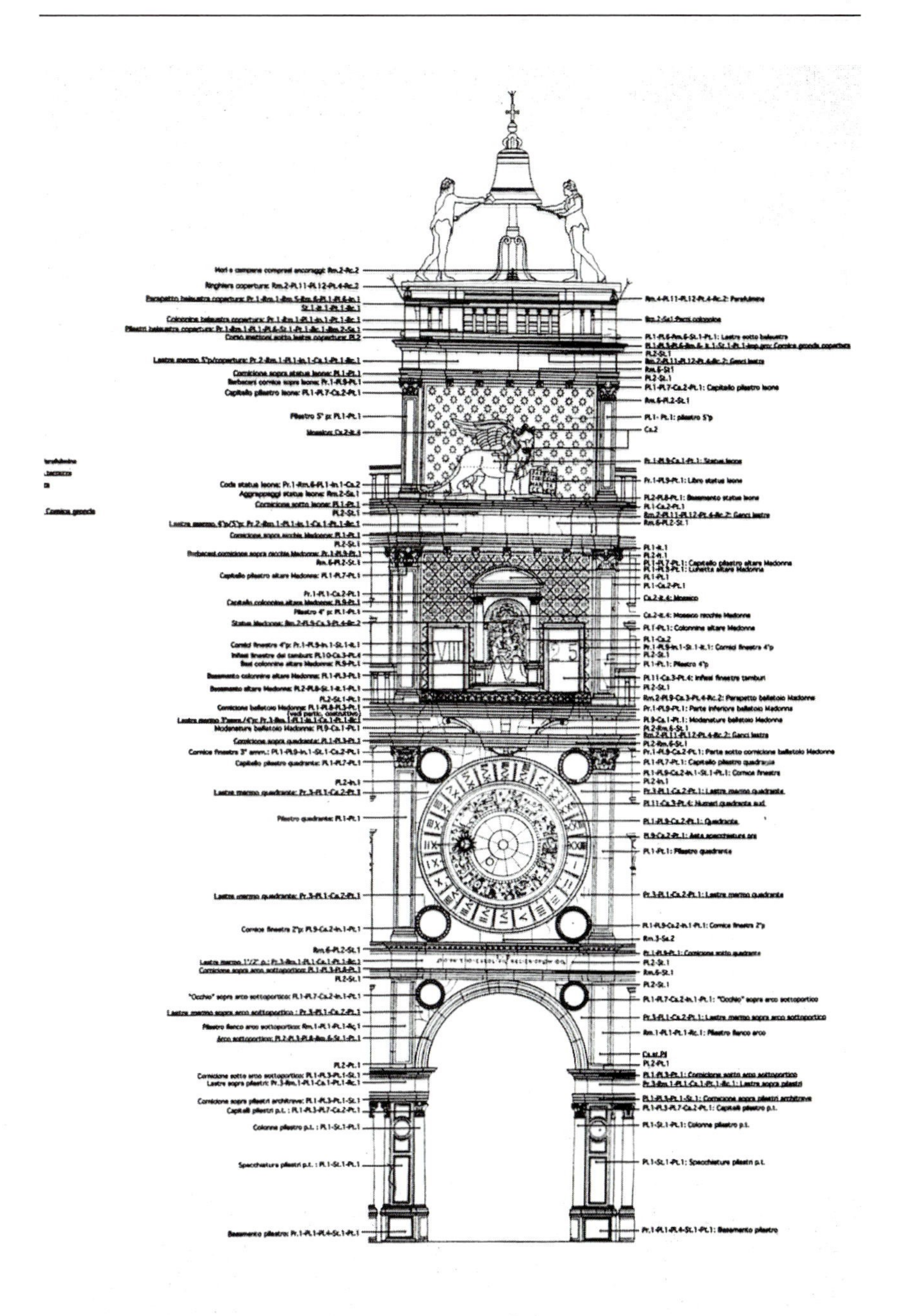

Fig. 9.5 The recent restoration design of the Clock Tower in St Mark's square (by G. Gianighian, M. Pandolfo, A. Torsello): main front.

floor walls, and the unjustifiable replacement of roof timbers and tiles. Control over such interventions is very sparse since and this kind of work is still, curiously, considered minor. The authorization permit is simple, controls over the execution of the work non-existent. The final inspection effected by the office, which releases the funds, is, by its nature, exclusively technical and administrative.

All this applies to buildings that are not protected and therefore not subject to the authority of the *Soprintendenza*, and which make up the overwhelming majority of cases. Such intervention does not conserve materials, but replaces them: it is carried out using materials and techniques intended for new construction. In the historic context the results are coarse and destructive. In conclusion, the public funds intended to conserve the historic built heritage of Venice are often expended on its destruction. Moreover, a useful indication of the relationship between re-use and destruction of historical materials can be gained by mapping the trends in the amount of rubble from building sites emptied into the public dumps of San Mattia in Murano. The figures are even more meaningful if one takes into account the fact that rubble from the most extensive developments has been dumped on the mainland. In 1988 the total quantity of waste unloaded was 71,815 cubic metres. Ten years later the amount had risen to 103,898 cubic metres, an increase of 31 per cent.

The value attached to historic objects, works of art or crafts objects, is not attached to those equally historic and much more widely diffused elements – the physical components of our historic buildings: the timbers of floors and roofs, the bricks and stones of our walls, the tiles on the roofs, the plasterwork, the ironwork, the staircases. Every one of these components and each individual element – each single brick, each single beam – has its own individual life and individual story that is worthy of attention and of respect. Such stories tell of places of origin 100 or more miles away: the woods and forests of the Dolomites, streams and rivers which transport the timber to the sawmills; islands in Istria in which stone is quarried and worked; mines in the Dolomites from which iron is extracted; furnaces in which were forged nails of a thousand different types – sheets, tie-rods, bars, hinges. These are the micro-histories to which our buildings bear witness, the palaces and the churches, but also the smallest of our dwellings. The real difficulty is to find a way to hear their story. And all of this, too, while taking into account the services essential to our contemporary lives.

The excavation of the canals

To this most rigorous interpretation of the conservation of the physical fabric of our historic heritage, we add another problem which in itself contains a series of problems, the excavation of the canals, which in turn entails the restoration of the foundations of the buildings facing them, and the maintenance/restoration of the network of streets. Let us examine these points one by one. We have already seen that the *Serenissima* took great care in keeping the canals of the city well cleaned in order to permit the free movement of the tides. This was never left unattended for any long period of time: at periodic intervals (between 10 and 20 years, according to the particular situation) each canal was drained, the mud dredged, the foundations repaired. It is clear that in this way regular control was maintained over the condition of the underwater structures, just as it was over the buildings and the streets facing the canals. This continuity, however, was interrupted, partly as a result of incompetent administration, partly because of practical difficulties. The pollution of the waters of the lagoon, and thus those of the canals, made it difficult to find a proper place to dispose of the material excavated from the canals, this became a good excuse for a lack of activity on the part of successive local administrations.

For 30 years, no canals were excavated, thus contributing to the general pollution of the waters and substantially raising the beds of the canals in a dangerous manner. In the case of very low tides, neither ambulances nor fire-fighting vessels could pass through many of the canals, with consequences easy to imagine and sadly proved by the fire of the Fenice.

Now that the *Comune* has both the means and the will to perform this work, as the administration of funds provided by the special laws has shown, these operations are already well advanced. Some difficulties, however, still remain, in part the result of the 'novelty' and the extent of this work. Various forms of intervention are carried out on the embankment walls, according to the degree of degradation and the state of the structure. They are in a very serious condition nearly everywhere, partly because of the long lack of maintenance, and partly as a result of the increase in *moto ondoso* (damaging waves caused by motorized vessels) (see below). The solution adopted most frequently is the substitution of the external faces of load-bearing walls, consolidating them with injections of 'microlime', into which steel reinforcing rods are inserted, to which a conventional electroplated steel reinforcement 'mat', with rods typically of 8 mm diameter is then fixed. Stainless steel (AISI 316) is not used, although it may appear more appropriate, because it is generally thought of as very expensive. In fact it would not be much more costly owing to the small quantities involved. The whole is then finished with concrete of a thickness varying from 6 to 10 cm.

The fate of inadequately thick reinforced concrete in a marine environment is well-known; such 'restoration', where executed, has also incidentally resulted in the loss of almost all of the outlets of the *gatoli* (ancient sewage-pipes, which conducted sewage into the canals), which is itself destined for rapid degradation. This method has, in addition, the further disadvantage of cancelling the individual structural elements, because the layer of concrete covers up an entire structure. Undoubtedly, the next restoration will require considerably more radical interventions, since it will be impossible to carry out timely intervention solely on the weakest parts of the structures, as was the case in the past. The next restoration will certainly also cost a fortune.

Venetian streets

Venice is an urban complex in which each element, however 'minor' it may appear, is a fundamental part of the creation of the image of the city: thus, the paving of Venetian streets is extremely important. Most paving consists of fine old stones called *masegni*: a *masegno* is a dressed piece of trachyte from the Euganean Hills, in the rough shape of a truncated pyramid, trimmed around the perimeter, and dressed on the visible face, while the other five sides remain rough. Dimensions vary, although the average thickness is around 15 cm. They are laid on a bed of sand, and in rows of the same breadth, with the straight joints in line with the principal route or direction of travel. The stones, bedded by means of pounding with a mallet, were butt-jointed with lime-mortar, almost touching each other. Laying the stones has to be almost perfect, particularly along the edges, in order to ensure minimum space between the *masegni*. One thus obtained a 'mosaic' pavement, the *tesserae* of which all worked together. The surfaces were also sloped to ensure that puddles could not form.

Today, the joints are much wider, with gaps of several centimetres, and filled with cement mortar. As is clearly apparent when it rains, the sloping has been forgotten, with consequent and evident inconvenience. But of greatest concern is the fact that many *masegni* disappear during the course of works, either because of incorrect specifications (as in the Campo San Giacomo dall'Orio, where all the old *masegni* were finished through sawing), or by carelessness in execution, demanded by speed. Another serious danger results from renovation works to services below the pavements (water, electric power, gas, telecommunications, sewage). Lack of co-ordination between the various utilities means that frequently the same tract of pavement is taken up two or three times in succession. Each intervention results in the destruction of many *masegni*. This fact has been the subject of a recent emergency appeal to the city council, claiming the loss of at least 11,000 *masegni*. It is certain that the total is much higher. The modernization of the telecommunication network with the installation of large diameter pipes under the ground and with the associated inspection-covers has resulted in numerous losses.

We are discussing here, both with respect to the paving and to the excavation of the canals, work that is of considerable cost and importance. However, the opportunity to survey existing conditions before works began was not taken, nor have strategies been developed that could direct the progress of this work.

Other problems

The present administration, although in many ways considerably more active and dynamic than those that preceded it, appears, however, interested above all in leaving a sign, a highly visible mark, of itself. The predominant emphasis is, in fact, on the new. To resolve the difficult and long-dated problem of lodgings with controlled rents, for example, the municipality favours the new construction projects to fill sites that are cleared or undeveloped, located at the outermost margins of the city. This sometimes results in demolitions that are not entirely acceptable. For example, in order to achieve the 'urban recovery' of an obsolete industrial site on the Giudecca (the former Junghans plant), it was decided to demolish one of the buildings, built in the 1940s, and hence not perhaps of exceptional value, but certainly unique of its type in Venice.

Large complexes, meant to house new university departments, are planned, together with new bridges over the Grand Canal, and much more besides (De Michaelis, 1999). This might all be very well, if plans were accompanied by proportional attention to the 'old', which is not the case. One is instead made aware of certain impatience regarding this type of problem, as if we were discussing something not so very important, or else too obvious and easily accomplished.

Environmental quality: the problem of pollution

Further dangers threaten the city and its lagoons, the most important of which is pollution. Venice suffers from its own source of pollution: the city has no comprehensive sewage network, and waste enters the waters, polluting and depositing mud in the canals. The result is a diminution in the amount of the sea-water that flows into the city and back to the sea twice a day, replenishing the canals and effecting the efficient, hygienic arrangement that, over the centuries, has constituted Venice's advantage over a majority of European cities. The principal agents of this pollution are the outlets of public buildings, of hotels and restaurants, of the few industrial productive processes that remain both in the city and on the mainland, and the run-off of agricultural fertilizers into the lagoon.

The laws against pollution have been in place for some time (the Law no. 206/95 requires that both the private and public sectors obtain authorization to discharge effluent waters into the canals). But it is only recently has the magistrature has begun to apply them, with increasing severity, imposing punitive sanctions that have caused a rush of remedial action to be taken on the part of the agencies. Today, there are numerous examples of installations for the treatment of sewage produced by tourist establishments of various types.

On the other hand, since 1974, the Special Laws passed at that time for the city have banned the use of oil and all other polluting fuel in heating systems, and this resulted in the dreadful '*macchia nera*' (black staining), a form of incurable cancer on the Istrian stone. All of these installations had to be converted to methane gas. This transformation was funded to a large extent with public money, and its implementation took place quite rapidly. In this manner, one of the gravest sources of pollution has been eliminated.

During the following years, and since the closing of the majority of the industries at Porto Marghera, (the industrial hub of Venice situated on the mainland) Venice has moved closer to a mono-culture based on tourism. This has resolved numerous problems and contributed to cleaner air and if we also take account of the fact that the city has no automobile traffic, the monuments of Venice are, from this point of view, better protected than elsewhere. Water traffic is certainly a source of pollution, but it is of a volume much smaller than that of any normal city. We have already noted the work of excavation of the canals, a programme of urban maintenance but which is also part of the civic environmental management.

Tourism and heritage management

Venice has therefore, over the last decades, in effect lost all economic and/or productive activity except that based on tourism, on which it now almost exclusively relies. The management of tourism is therefore a primary factor in the life of the city, and approaches to conservation cannot ignore it. It is notable that in the last 20 to 30 years, so-called 'mass tourism' has affected those cities blessed with an important artistic and cultural heritage on a massive scale, i.e. those cities which have ultimately been defined as 'art cities'. In many of them, the heavy pressure of visitors has by now filled to maximum capacity the tourist facilities and accommodation in the city, so as to put at risk the social–economic balance, the quality of life, the residents' well-being, even the physical survival of the affected city centre itself. Thus, in the end, instead of helping the development of the city, tourism risks killing it; in essence, the pressures of tourism are no longer sustainable. This is true for many art cities, and particularly for fragile Venice.

The concept of 'mass tourist load' or capacity of tourist reception has been analysed (Canestrelli and Costa, 1993). Experts have fixed 25,000 visitors per day (14,000 of whom are defined as 'excursionists', that is, those who do not spend even one night in the city), as the critical limit beyond which the influx of tourists becomes unsustainable. The limit is already surpassed for at least half of the year, and it has been forecast that it will be exceeded on 216 days a year by 2000.

Another indicator that quantifies the pressures of tourism and the inconvenience it generates is the relationship between the number of visitors, the number of available beds, and the number of the city's inhabitants (the latter, as we have seen, significantly reduced over the last decades). If we look at these indices together with those of other major tourist centres, we find that Venice faces a higher proportion of visitors than the others. Considering solely the population of the historic centre, excluding Mestre, which absorbs no significant tourist flows, there are 90 tourists per year for each resident, compared to 36 in Salzburg, 23 in Bruges and 10 in Florence. Moreover, the number of available beds in proportion to the population is 15 per inhabitant in the historic centre, against 7.3 in Salzburg and 6.7 in Bruges (Van der Borg and Russo, 1997). Venice is therefore in a more difficult situation, particularly regarding access and transport. Nevertheless, although numerous studies have been undertaken to assess the political possibilities of controlling the influx of tourists, no strategy has so far been developed to control and manage them.

In discussing the management of the influx of tourists, and of access to the city, we must not neglect the problem of the planning of terminals for the city, particularly those relating to motor vehicles. It should be emphasized that the terminal for vehicular traffic in Venice is also the point of interchange from the vehicular to the water system, with problems, therefore, rather different from those simply related to parking.

At present there are vehicle terminals at Fusina, San Giuliano, on the island of Tronchetto and Piazzale Roma. The latter is the most important, since it is the closest and most accessible to and from the historic centre. It is the point of arrival of the bridge, which links Venice with the mainland, and is the route of motor traffic both public and private, and also for pedestrians. This zone is an open wound, which successive administrations have modified from time to time, without ever effecting a satisfactory cure.

Pedestrian and vehicular movements are inextricably tangled up creating, at times, traffic jams that would not be out of place in London or Paris. The Ponte della Libertà on these occasions is often blocked by traffic, cutting the city off from the mainland, and completely preventing any movement. In addition, when there is an accident on the bridge, there is no possibility of diverting traffic onto the lanes for the other direction, as is usually done on motorways; it simply becomes impossible, for a certain number of hours, to cross the bridge in that direction. The difficulties this situation creates are easy to imagine. This, too, is part of the overall problem the historic centre imposes on those who still live there, as well as on those who wish to visit.

Like many other historic centres, too, Venice is suffering from the loss of shops serving the needs of the residents (general food shops above all, but also drapers, stationers, bookshops and similar establishments), in favour of shops serving the tourists. In the city centre, the baker is replaced by Armani, the butcher by Gucci, while everywhere in the peripheral districts there is a flowering of shops selling 'Venetian specialities' – which consist mainly of small glass objects and masks, the production of which is recent but has been an immediate success and is now widespread.

A brief mention of the Jubilee year and the matters of finance associated with it is necessary. The total cost of the planned operations is put at Lit. 63 millions (approx. £22m, US $35m), an expenditure almost entirely allocated to work in relation to low cost accommodation. This has given rise to often brutal 'restoration' work on historic religious or quasi-religious complexes such as old monasteries and nunneries that, today, are almost all unused. These have been completely gutted and restructured, destroying their old roof tiles, ironwork and so on, much of which was in a condition that could have been restored. Once again with little or no control over the quality of the work, as a very short period of time was allowed for any proposed project, as a condition of obtaining funding. It seems, too, inopportune to increase further the tourism accommodations in a city which to date has not been able to produce any policy for a tourism management to make it compatible with the life of the city, nor of controlling the masses of tourists that already crowd its streets.

The city is subject to yet another stress: that of the *moto ondoso*, damage from the wash of motor boats. We refer here to a well-known phenomenon, although one that has not been precisely quantified, but which has put at serious risk, over a short period of time, the very survival of buildings located on the Grand Canal, as well as those on the minor canals. Attention was brought to the fact that these vessels 'displace a mass of water and create vortices which attack the lower foundations and produce an intermittent suction effect on those foundations, stirring every fibre of the underwater structure of the city' more than 30 years ago (Gazzola, 1967). These motor boats are not decreasing either in their power or in their numbers. Other than the pedestrian network, water traffic is the only means of transport in Venice. As many as 15,000 vessels today circulate in the extensive network of canals both narrow and broad, within the city and in the surrounding lagoon. A good 90 per cent of this fleet consists of transport boats of various types, from the oar-powered *sandoleto* (small traditional Venetian boat) which accounts for around 105 of the total, to the luxurious 30-metre yacht. (The traditional Venetian Laguna vessels, powered solely by oar or sail, accounted for the majority of the total 50 years ago.) The percentage of vessels serving several functions of general interest, transporting goods and passengers, accounts for around 10 per cent of the total, that is, about 1,500 vessels, about half of which are used for the delivery of goods or other commercial uses, while the other half serve passengers, residents and tourists.

The public transport fleet consists of approximately 150 vessels, serving the waterbus routes which total 167 km in length; this fleet transports 14.5 million passengers per year. A further 600 vessels comprise the private fleet, and consist, above all, of water taxis (a source of great gravity: taxi drivers constantly blackmail the local authorities and appear to be quite uncontrollable), flanked in ever increasing numbers by large collective passenger vessels, used solely to serve tourism. The mass of traffic is therefore daunting, and at present no realistic means to control it are apparent.

At the terminals, both for the railway and for motor vehicles, an average of 80–90,000 people (workers, students, tourists) converge every day, the arrival of whom doubles the resident population of the historic centre, and creates daily peaks of demand on the transport system. This is probably one of the points on which the very survival of the city will depend: transport is essential, and in the view of many it is not sufficiently rapid to permit the development of the normal activities of a living urban centre. However, this proposition perhaps should be doubted: bearing in mind the insoluble parking problems and the ill-fated traffic blockages that are daily inflicted on 'normal' cities, rendering complex strategies concerning hours and routes necessary for those wishing to use private transport, the public systems of transport, at least in Italy, are frequently quite inadequate.

Conclusions

From this study of Venice, that incomparable historic city – an absolute metaphor for the world's heritage, a concentration of its values and history – it is clear that unsolved problems and risks remain today. In part these are the result of its own fascinating attractiveness, and in part are caused by an administration incapable of resolving problems inherited from decades of superficial or irresponsible political management.

The policies that determine planning have not addressed at its root the most critical nexus of the city, which is that of *conservation* – the need to sustain the historic fabric. The present administration, more responsible than the previous ones, still does not seem to comprehend that the absolute priority in the city is the preservation of its historic heritage. Its priority, therefore, becomes the encouragement of smart new architecture, trying through this approach to slow the population exodus, while, at best, the management of what is historic is left to the control of bodies that are purely bureaucratic. Often there is no control at all (as in the examples given here of the loss of plasterwork, bricks, roof tiles, *masegni*, ancient drainage systems and so on). One has almost the impression that 'regenerative' policies for the preservation of the historic centre do not really exist, other than as a desire to leave signs of 'the modern' in various peripheral areas, at present obsolete. The public administration disburses large funds, but these funds damage conservation as much as assist it. The conservation work is exclusively concentrated on the great monuments, such as the Ca' Pesaro and the Ca' Rezzonico, but the living texture of Venice's constructed fabric will soon be completely lost unless provision for its rescue is made a matter of high priority.

A separate discussion needs to take place regarding the work of the *Soprintendenza* of Venice. The interventions of this state-run organization are always of a very high quality, conducted with great skill and judgement, based on refined theoretical principles, such as to constitute models for appropriate conservation procedures. The restoration of the façade of the basilica of San Marco (Porta di Sant'Alipio) has been cited by many as an exemplar, particularly in contrast with the very recent discussions over the cleaning of the façade of San Pietro in Rome. We should also mention, as another example, the restoration of the façade of the church of Santa Maria del Giglio (Fig. 9.6), which has brought back to the building a forgotten splendour, without compromising any of its various historic elements. This

Fig. 9.6 The façade of the church Santa Maria del Giglio. (Photograph: M. Fresa.)

good practice is also seen in the projects funded by international agencies, such as 'Venice in Peril', and carried out in collaboration with the *Soprintendenza*. These agencies also do valuable work in heightening international awareness of the problems facing Venice and its urban tissue.

Furthermore, the *Soprintendenza*, to which falls the control of all operations that take place on protected buildings, also exercises a commendable job of supervision and education about conservation practice, indicating the criteria and methodologies that should be adopted by professionals, both public and private. However, this body is primarily concerned with those buildings which are the property of the state, and its authority extends only to those buildings that are subject to legal protection, numbering only around 1,500 in the entire *Comune* of Venice.

Although in principle highly positive, the interventions on the canals (Fig. 9.7) and foundations also appear problematical, since they are carried out using highly destructive procedures. The same consideration applies even more tragically to the technical operations that have devastated the ancient pavements and removed much historic material from below the level of the ground. Thus, to give just one example, during the work to the

Fig. 9.7 Removing mud from the rii (canals).

sewage system, not far from the church of the Frari, a large *gatolo* was seriously and pointlessly damaged, nor did anybody try to stop the job. The point is that there does not seem to be any legal way to stop this kind of work: no one seems to have authority under the ground, except in cases of archaeological findings. Only a general public sensibility to given values could produce juridical innovations, but this conscience is still incapable of grasping the importance of historic material *per se*. We still tend to judge on the basis of vague aesthetic concepts (it is beautiful – it is ugly) and of the relevance of a built structure. Thus has entered into common awareness, and also into legislation, the perception that monuments are to be conserved within their environment, that the ancient city is a *unicum* to be preserved as far as possible intact. The principle that one should not all but demolish a building, leaving intact only the exterior (*façadism*, today undergoing high criticism from the experts) is slowly growing in importance, while the perception of the value of historic materials, even the humble, 'non-artistic' ones, is still shared by only a few. This attitude is considered an almost crazy, fetishist approach to the built heritage: many battles are still to be fought on this topic.

If we then pass on to examine the problem of the sustainability of tourism and its management, the issue becomes still more serious. In this field, in fact, the administration has not developed an acceptable policy for managing tourist flows, so that even at present, and even more so in the future, tourism, instead of helping the city, will become its enemy. Tourism will constitute a daily increasing danger, not only to the quality of life of the residents but also to the general well-being, both physical and social, of the city. This is a problem that today affects the large majority of historic centres. Excessive tourism, together with the loss of the resident population, is now such as to put at risk the very survival of the city, and is today one of the most difficult problems for the 'art cities'. The city must survive without transforming itself into a cultural Disneyland; at the same time we should not think of imposing limitations that are too rigid, which would impede the legitimate and universal appreciation of this most precious and so fragile and delicate a heritage.

On the solutions to these issues depends the future of Venice, as does that of the other cities of Italy (and not just in Italy), all of which carry the burden of the history that constitutes our most priceless heritage.

References

Beltrami, D. (1954): Storia della polazione di Venezia dalla fine del secolo XVI alla caduta della Repubblica, Padova.

Braudel, F. (1979): *Civilisation matérielle, économie et capitalisme*, XVe–XVIIIe siècle, 3 vol., Paris.

Canestrelli, E. and Costa, P. (1993): 'La determinazione della capacità di accoglienza turistica, un approccio sfocato', in Costa, P. (ed.) *Venezia. Economia e analisi urbana*, Milan, pp. 269–88.

Choay, F. (1988) : *L'allegorie du patrimoine*, Parigi.

Comune di Venezia (a c. Settore Coordinamento Legge Speciale) (1998): *Interventi per la salvaguardia di Venezia e la sua laguna*. Programma e stato di avanzamento

(Leggi: 5 Febbraio 1992 n. 139, 20 Dicembre 1995 n. 539, 4 Ottobre 1996 n. 515, 2 Ottobre 1997 n. 345), pubblicazione periodica aggiornata al 31.12.1998.

Connell, S. (1988): *The Employment of Sculptors and Stonemasons in Venice in the Fifteenth Century*, Garland Publishing, Inc., New York & London.

Crouzet Pavan E. (1992): *Sopra le acque salse: Espaces, pouvoir et société a Venise*, 2 vols, Rome.

De Michelis, M. (a c. di) (1999): *Venezia. La Nuova Architettura*, Milan.

Gazzola, P. (1967): 'Venezia in pericolo', in *Per la salvezza dei beni culturali in Italia*, vol. II, Rome, pp. 605–9.

Gianighian, G. (1989): 'La casa complessa del Rinascimento', in Maire-Vigueur, J.C. (ed.), *D'une ville à l'autre: structures matérielles et organisation de l'espace dans les villes européennes (XIII–XVIe siécle)*, Rome, pp. 557–90.

Gianighian, G. (1990): 'Appunti per una storia del cantiere a Venezia (secoli XVI–XVIII)', in Caniato, G and Dal Borgo, M. (eds.) *Le arti edili a Venezia*, Venezia, pp. 237–56.

Lorenzetti, G. (1974): *Venezia e il suo estuario*, Trieste.

Pavanini, P. (1989): 'Bonifiche urbane nel XVI secolo a Venezia', in Maire-Vigueur, J.C. (ed.) *D'une ville à l'autre: structures matérielles et organisation de l'espace dans les villes européennes (XIII-XVIe siécle)*, Rome, pp. 485–507.

Pertot, G. (1988): *Venezia 'restaurata'. Centosettanta anni di restauro sugli edifici veneziani*, Milano.

Romanelli, G.D. (1988): *Venezia Ottocento: l'architettura, l'urbanistica*, Venezia.

Trincanato, E.R. (1971): *Venise aul fil du temps*.

Van der Borg, J. and Russo, A.P (1997): *Un sistema di indicatori per lo sviluppo turistico sostenibile a Venezia*, Fondazione Enrico Mattei, Progetto Venezia 21, Venezia.

Wirobisz, A. (1965): 'L'attività edilizia a Venezia nel XIV e XV secolo', in *Studi Veneziani*, VII, pp. 307–43.

Juris Dambis

The historic centre of Riga, Latvia

Introduction

In comparison with Western Europe, Latvia is rather sparsely inhabited. Cities are smaller and they are located far from each other. Each city has its significance and its own peculiar visage (Fig. 10.1).

Each region of a city is defined by several factors that later affect its development. Historical investigations show that at the end of the first millennium and in the beginning of the second millennium peasants, artisans, and merchants inhabited localities formed around ancient Latvian castles – ancient towns. They mainly originated by the side of water and overland strategic routes. Until the sixteenth century there were 11 towns in Latvia, up until the seventeenth century, 10 towns, and up to the nineteenth, 5 towns; now there are 77 towns, 8 of which form part of the present republic.

The largest of the cities and the richest in history is Riga, founded in 1201. It has 805,997 inhabitants at present; the smallest town, Durbe, was founded in 1893 and it has 465 inhabitants at present. A peculiarity of Latvia is that one third of all the inhabitants of the state live in the capital Riga.

Riga has always been the largest city in the Baltic territories – a centre of industry, trade, science and culture (Fig. 10.2). Until the middle of the nineteenth century it was a fortress enclosed by ramparts, fortified walls and moats. In 1867 the ramparts were removed and the surrounding area was developed according to a master plan. What had once been a fortified esplanade was replaced by a semi-circle of boulevards enclosing the old centre of Riga and a grid-iron pattern of suburbs was developed beyond these boulevards. It is here that the new buildings in the style of Art Nouveau, for which Riga is famous, were developed.

After the restoration of independence in Latvia, the juridical and economical independence of the municipalities was restored following the law

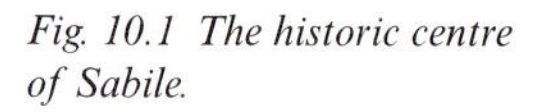

Fig. 10.1 The historic centre of Sabile.

Fig. 10.2 The silhouette of Riga.

'On Town Municipality' passed in 1991. At present 68 per cent of Latvia's inhabitants live in towns. At the beginning of the twentieth century, interest began to arise in the preservation of historic town centres in Latvia; however, earnest practical action began in the year of 1967 when Old Riga was proclaimed a state 'protected zone'. From 1977 until 1983 the first regeneration project was worked out; before that an investigation of several historic towns was commenced. At present, there are 44 town-planning monuments under state protection in Latvia.

On 4 December 1997, the historic centre of Riga was included in the UNESCO World Heritage List due to its 'outstanding universal value'. The medieval and later urban fabric remains relatively intact; however, the quality of its Art Nouveau architecture, which is unparalleled anywhere in the world, and its nineteenth-century timber architecture, were the reasons for meriting this 'outstanding' status (Fig. 10.3).

The policy and planning framework

The law on Territorial Development Planning (passed in 1998) established that municipalities are responsible for ensuring the production of territorial development plans. This law defines the principles and assignments for territorial development planning, the order of public deliberation. The law defines the areas of responsibility in the field of planning for different state institutions and municipalities. The aim of developing any territorial development plan is to ensure the following principle:

> A principle of a sustainable development which provides for ensuring a qualitative environment, a balanced economical development, a rational use of nature and human material resources, and a preservation of cultural heritage for future generations.

Historic town centre planning policy is based on ensuring the development of a town that respects the cultural heritage and uses it as an economic asset. History embodied in cultural monuments ensures a balanced, humanized and valuable environment for human life. Objects related to the history of civilization are regarded as being the cultural, intellectual, social, and economic capital of society, and therefore it is an irreplaceable value. This capital forms itself over the course of many centuries and destruction

Fig. 10.3 The historic centre of Riga.

of any part of it makes the historic centre of a town become poorer. Cultural heritage has to be handed over to future generations in an authentic condition. Each period has its features and layers that are also historical witnesses and have cultural significance. All new transformations have to respect the original values; although they should also present the features of their time. Creation of false images in the historic centres of towns is no way to preserve the cultural heritage – this approach should be regarded as being negative (Fig. 10.4).

It is necessary to work out a preservation and development project for the historic centres by using existing general plans. The objective is to create an effective tool for the control of changes:

- to control preservation and construction in the centres, ensuring a rational use of the particular territory which would promote the development of a balanced economy and preserve the environment (including the cultural environment) corresponding to the state and municipal sustainable development priorities and development plans;
- to ensure a qualitative housing environment for individuals and society in general;
- to guarantee the rights of land proprietors and users to utilize and develop their property or the land utilized in accordance with the terms of territorial planning;
- to provide society with opportunities to acquire information, express one's opinion and take part in the process of the development of the territorial planning.

The most significant constituent parts of a project for managing historic centres in Latvia are:

Fig. 10.4 The reconstruction of lost housing in the form of 'copies' is not the most successful method.

- an inventory and an analysis of the present housing schemes;
- analysis of the town planning situation;
- a plan indicating the borders of plots;
- analysis of
  - underground (archaeological) cultural and historical values,
  - the transport system,
  - communication facilities,
  - the use of territory and constructions (by zones),
  - the existing greenery,
  - existing infrastructure,
  - pollution,
  - employment,
  - the guidance and control system over preservation and development of the historic centre;
- proposals for town planning and development;
- solutions for transport and traffic and car parks;
- development of communications;
- perspective and functional use of constructions (by zones);
- details of housing schemes (sizes);
- identification of streets (by red lines);
- development of the greenery;
- development of gutters and water containers;
- instructions for the preservation of cultural and historical values of each block;
- instructions for the development of housing schemes;
- development of infrastructure;
- a mechanism for modification, control and administration;
- involvement of the society in the creation of a qualitative environment and availability of information.

The registration and evaluation of the existing qualities are the most important activities to make a start in land-use planning. The investigation of unified housing types within historic centres in Latvia has been commenced, as well as the registration of other types of construction.

This investigation is carried out by identifying the following issues:

1 *The address of the building.*
2 *The proprietor of the building:* a private, state, or municipal property.
3 *The status of the building:* a monument of state, or local significance, whether it is a significant constituent part of the town-planning monument, or whether it is an object degrading the environment.
4 *The period when the building was erected:* from the twelfth century to the end of the seventeenth century, from the eighteenth century to the nineteenth century, from the middle of the nineteenth century to the year of 1914, from 1920 to 1940, from 1945 to 1990, and after 1990.
5 *The primary use of the building:* dwelling, public (social), or household (industrial) building.

6 *The architectural style of the building:* Gothic, Renaissance, Baroque, Classicism, Eclecticism, Art Nouveau, Neo-classicism, Functionalism, Neo-Eclecticism, architecture after the year of 1954, or a building without a style.
7 *The use of the building:* the ground floor, other floors, basement, attic; public rooms, flats, or not used.
8 *The number of the storeys* (above the ground).
9 *The material and type of construction:* the walls: brick, combined; the roofing: tile, tin, natural slate, concrete, ruberoide, other.
10 *The windows of the main façade, authenticity:* all are original, more than a half, less than a half.
11 *The windows of the main façade, by the (predominant) material:* wooden, plastic, metal, whether there is stained-glass in the windows of the staircase.
12 *The outer doors of the main façade, authenticity:* all are original, more than a half, less than a half.
13 *The outer doors of the main façade, by the (predominant) material:* wooden, plastic, metal, other materials, stained-glass, grindings.
14 *The transoms of the outer doors of the main façade, authenticity:* all are original, more than a half, less than a half.
15 *The transoms of the outer doors of the main façade, by the (predominant) material:* wooden, plastic, metal, other materials, stained-glass, grindings.
16 *The finish of the building's façade:* plaster, bricks, natural stone, tiles, artificial stone plates, wood, relief, sculptures, other materials.
17 *The interior decoration of the openly accessible inner rooms:* whether of especially high artistic value, of high artistic value, of no special value.
18 *The evaluation of the building's technical condition on a 10-point system:* 1 – ruins, 2 – breakdown, 4 – poor, 6 – satisfactory, 8 – good, 10 – excellent.
19 *The necessary proceedings for the salvation or repair of the following parts of the building:* foundation, walls, roof constructions, roofing, cornices, gutters, balconies, decoration of the façade, doors, windows, other elements.
20 *The role of the building in the historic centre:* whether the building is unique for its content; the building site is significant for the scenery of the street; the building is the creation of an outstanding architect; the building is a closing of street perspective; the decoration is extremely valuable, richly detailed, or the building is a corner building.
21 *A picture of the building:* the film number/the negative number.
22 *Information on the visible recent repairs; yes or no:* to confirm, the inventory case information of the particular building is available for inspection.
23 *Information on visible damage caused by reconstruction; building elements that have degraded the environment of the historic centre: yes or no:* to confirm, the inventory case information of the particular building is available for inspection.

Management and regeneration action

In Latvia, municipalities are responsible for the development of the territories. Special services have been formed within them. These are known as 'Building Commissions'. Without the permission of the Building Commission, no construction or reconstruction of existing buildings is permitted. In conformity with the law *On Preservation of Cultural Monuments*, state administration over the protection and use of cultural monuments is provided by Cabinet of Ministers, and is carried out by the State Inspection of Heritage Protection.

Some municipalities have established their own cultural heritage services – principally in the largest and culturally most significant historic town centres. To define the areas of responsibility, the Inspection and the municipality enter into a contract (an agreement). In order to improve the co-ordination among the co-operation of the institutions interested, a 'Cultural Heritage Development Co-ordination Board' is formed to co-ordinate action for the historic centre. It includes responsible officials both from state and municipal institutions, as well as from the most significant professional and public institutions.

Building regulations are issued for towns, which also include the requirements for the preservation of cultural heritage. In cases where the requirements are ineffective, the Inspection service can issue regulations for the preservation of the historic centre, which are binding for plots and building proprietors.

'Instructions' on the use, reconstruction and restoration of the historic centre of Riga have been issued in addition to the building regulations and remain in operation until the preservation and development project of a composite historic centre is worked out. The regulations are based on article 5 of the law On Protection of Cultural Monuments.

The execution of these 'instructions' should guarantee the agreement that the Republic of Latvia entered into when it became party to the *Convention on Protection of World Cultural and Nature Heritage* (in particular: article 4 of the convention).

Action for the historic centre of Riga

Since December 1995, when the historic centre of Riga was first put forward as a candidate for inclusion in the UNESCO World Heritage List, a programme of activities has been developed to ensure the protection of this World Heritage Site.

In 1997, the historic centre of Riga was inscribed in the World Heritage List and a 'Cultural Heritage Development Co-ordination Board' was established. The Co-ordination Board was comprised of representatives of the state and municipal institutions and representatives from UNESCO. Its main objective was to find solutions to matters concerning the preservation and development of the historic centre.

In 1998 the Co-ordination Board decided to devise a programme of conservation and development and took action to record the buildings in the historic centre which resulted in 4,000 buildings being investigated following work by the State Inspection for Heritage Protection. Further action was subsequently taken by the State Inspection, in association with the Norwegian National Directorate for Cultural Heritage, which resulted in the publication of practical maintenance guidance for owners of buildings in the historic centre.

During 1999, the State Inspection, in association with the State Land Cadastre Centre, developed a computer program to assist in controlling conservation work and new development in the centre. Further actions in 1999 included the preparation and adaptation of binding guidelines (the 'instructions') for the maintenance, reconstruction and restoration of housing; the opening of an Information Centre by the State Inspection; and a professional workshop, organized by the Cultural Heritage Department of the Council of Europe, to assist in the development of a 'Rehabilitation Plan' for the centre.

Other actions will be developed in 2000, and 2001 – when Riga will celebrate its eight hundredth anniversary. These include: the formulation of a 'Preservation and Development Plan'; the completion of the inventory of buildings in the historic centre; a work programme for the conservation of the timber architecture of Riga; and will culminate in an international conference organized under the auspices of Europa Nostra in conjunction with the official opening of the 'European Heritage Days' in 2001.

The organization and management of all these activities is being undertaken by the Co-ordination Board and is being sponsored by Riga city authority, the state budget, other European countries, international organizations and other sponsors. This high profile action reflects the international importance of the historic centre of Riga.

The 'instructions' used for the historic centre of Riga

The 'instructions' are now used to control the use, repair, conservation, reconstruction, and restoration of existing buildings, as well as the erection of new buildings within the defined territory of the historic centre of Riga – the territory of the UNESCO World Heritage Site No. 603. In conformity with article 5 of the law On Preservation of Cultural Monuments, the instructions are binding upon all the proprietors and directors of plots, buildings, and constructions that are located in this area. This World Heritage Site includes the territory restricted by: E. Melngaiļa Street, Kr.Valdemāra Street, Palīdzības Street, Blieķu Street, Tallinas Street, A. Čaka Street, Matīsa Street, Avotu Street, Lāčplēša Street, E. Birznieka-Upīša Street, Elizabetes Street, Satekles Street, Marijas Street, Gogoļa Street, Turgeņeva Street, 11 Novembra Embankment, Muitas Street, Citadeles Street, the city channel and Eksporta Street.

The historic centre of Riga includes an archaeological monument of state significance – No. 2070 – the 'Archaeological Complex of Old Riga' (see below), the borders of which include the territory restricted by: 11 Novembra Embankment, Kr.Valdemāra Street, Aspazijas and Basteja Boulevards, 13 Janvāra Street. It is also a constituent part of a town-planning monument of state significance – No. 7442 – the 'Historic Centre of Riga City'.

The recognized and preservable values of this centre include the structure, planning and construction form of historic buildings; the dimensional disposition; the scenery, panorama and silhouette; the greenery system; the construction of places and streets and their dimensional disposition; the dimensional organization of blocks (organization of public services and facilities); elements of improvement; materials of inner decking; fences, fountains, historical constructions and their remains in the cultural layer; characteristic relief and waters.

For each cultural monument of state or local significance within the historic core the State Inspection of Heritage Protection has provided instructions on permissible uses and preservations requirements. These instructions were co-ordinated and reviewed with the assistance of the Co-ordinating Board at the beginning of 1999 and cover the following issues.

Cultural layer
Within the archaeological complex of old Riga all kinds of groundwork, except for the repair of the existing engineering communications, have to be confirmed by the Riga Inspection of Heritage Protection. To begin the work, written permission is required from the State Inspection. Historical constructions and their remains uncovered during archaeological investigations can be disassembled or taken down only if approved by the State Inspection.

Reconstruction of blocks and streets, places and squares
Where new large buildings are proposed along with any reconstruction of the adjacent blocks, a 'detailed plan' of the particular territory is obligatory. Reconstruction of public outer spaces (streets, places) are only admissible when a detailed plan has been worked out and has been confirmed by the authorities following the order established by law. A precondition for framing a detailed plan is that regulations have been worked out by Riga Inspection and approved by the State Inspection. Proposals involving the reconstruction of the building's size and silhouette are reviewed by the Coordination Board – this is a requirement in accordance with precept No. 142 issued on 11 November 1998 by the Ministry of Culture, and conformed in the regulations of the State Inspection.

Silhouette and panorama
To preserve and safeguard the silhouette and the panorama of the historic centre of Riga, the best circumstances must be found. The optimal zone for perceiving the silhouette and panorama is the left bank of the River Daugava from the railway bridge to Loču Street in Ķīpsala.

Roofs
Roofs should be regarded as a significant element of the historic centre of Riga. The original shape of roofs should be preserved. It is prohibited to replace the historical inner coverings of the roofs with uncharacteristic materials. When replacing a covering, historical materials are preferred – according to the architectural style of the building (natural slate, tile, painted tin).

Repair, restoration and replacement of façades, windows and doors
It is prohibited to replace the characteristic, original (historical) decoration of façades. Replacement of original wooden and metal windows and doors of the main (street, place) façades in the historic centre of Riga is prohibited. Windows and doors of an original construction of the main (street, place) façades have to be preserved and restored (Fig. 10.5). Their replacement with modern windows and doors is admissible only in the façades onto the courtyards (Fig. 10.6).

Exception can be made in relation to buildings erected after 1954; a general replacement of decoration, windows and doors in the façades is admissible. Original windows and doors of buildings erected after 1954 that have been replaced with ones irrelevant to the style, can be replaced with new ones on the condition that they correspond with the style of the building in division, shape and details. It is prohibited to install telecommunication appliances and digital code keys in the door casements.

Disposition of advertising elements on façades and backfire walls and satellite antennas
It is prohibited to place such advertising objects, as well as satellite antennas, on the buildings of the historic centre of Riga in a manner that would alter their façades or silhouette. Open backfire walls that are partly or fully exhibited to the street may be used for placing advertising objects if they are relevant to the character of the building's façades; however, illusory architectural motives or historic advertising traditions must be used. The colours of the advertising object must match tonally with the

Fig. 10.5 Original windows occupy a significant place in the scenery of the city's streets.

Fig. 10.6 Preservation of the original portals is important.

colours of the building's façades. The illustrative and informative block of the advertisement placed on the backfire wall in a compact form must not occupy more than 15 per cent of the surface of the backfire wall exhibited to the street. Only one advertising object may be placed on every backfire wall.

Transformation of ground floors

Ground floors comprise street space most actively. The same regulations as to the building in general are applied to the entrances, portals, show-cases, marquises (sunshades) and details (metal forging, flag holders, etc.) of the buildings in the historic centre of Riga. It is prohibited to cover window and door openings and show-cases with impenetrable blinds. It is prohibited to remove layers on façades that characterize a particular period of time (informative, remembrance, signs, etc.) without the permission of Riga Inspection.

Decking, illumination, greenery small architectonic forms

It is prohibited to replace street and place profile or to remove the roadway decking or its details. The replacement of roadway decking with another material in the streets and places of historic centre of Riga is prohibited. Preservation of the historic scenery and inner yards is important (Fig. 10.7). Street illumination must be differentiated as it depends in which part of the historic centre of Riga it should be set up. Artificial illumination of architecturally inferior buildings must not dominate over the illumination of the adjacent buildings that are of higher value. At night it is admissible to cover show-cases with grates which are of no less than 60 per cent lucidity.

Fig. 10.7 Preservation of the historic scenery and inner yards is important.

Destruction of objects
It is prohibited to destroy buildings, constructions and other objects in the historic centre of Riga, except for inferior objects or objects degrading the environment; such objects can be destroyed if confirmed by the State Inspection. Applications stating that the intention of the proprietor is to destroy a part or the whole object must be submitted to Riga Inspection and reviewed by the Co-ordinating Board. In addition, the following information must be added to the application:

- an architectonic and artistic investigation (for the objects erected before the year of 1850) or an architectonic and artistic inventory (for the rest of the objects) that includes a land survey;
- a conclusion on the technical condition of the object provided by a building engineer or architect certified in the field of restoration;
- an assessment provided by a certified architect on the changes in the historical and cultural environment that may occur as a result of the destruction.

Copies of the above mentioned documents are submitted to the State Inspection and Riga Inspection for permanent storage. In case the specialists are suspected of being incompetent, the State Inspection can demand a reiterative conclusion from other experts. Destruction of the object must be done in accordance with the regulations. Without the confirmation from the State Inspection, the permission to destroy an object is not valid.

Co-ordination of projects
Co-ordination of the transformation, repair, conservation, reconstruction, and restoration of buildings, constructions and other objects, as well as erection of new constructions and their projects in the historic centre of Riga, are carried out as defined in the normative acts and laws. The co-ordination must include the following information: the name of the object, its address, group, ground, the title of the project, stage, designer, customer, co-ordination date, number and the signature of the person responsible for the co-ordination. If the co-ordination lacks any of these positions, it is considered not to be valid.

Receipt of permissions and inception of works
To begin a transformation, repair, conservation, reconstruction and restoration of a building, construction or other objects, as well as an erection of new objects in the historic centre of Riga, one must receive printed permission of a special type issued by the State Inspection.

Environmental management

The law on environmental protection in Latvia was passed on 6 August 1991 with amendments made on 5 May 1997. The aim of the law is to create such a social and natural interaction mechanism to guarantee environmental protection, effective nature management and inhabitants' rights to have a qualitative living environment. The basic principles of environmental protection are:

- provision of a friendly living environment in the historic centres of towns for the general living, work and relaxation with respect to both present and future generations;
- co-ordination of ecological and economical interests of the society;
- provision of environmental protection measures,

The Ministry of Environmental Protection and Regional Development, its local services, and the environmental protection services of the municipalities, are responsible for environmental protection in Latvia. Action is being co-ordinated with the State Inspection in this respect for Riga and elsewhere in the country.

Tourism and heritage management

Cultural heritage in Latvia is a significant tourism resource. The Latvian law on tourism was passed on 17 September 1998. Through this law a juridical foundation for tourism development has been created. The law states the order according to which state authorities, institutions, municipalities and tourism enterprises have to work in the field of tourism and protect the rights of the tourists.

One of the main assignments of this law is to encourage the preservation and use of the cultural, historical and natural heritage, as well as to ensure the development of the cultural and natural tourism. In this respect Riga has an important role to play due to its importance as a World Heritage Site.

According to the law, municipalities are required to work out development plans and territorial planning requirements, defining the perspective of the tourism development; according to the development plan and territorial planning, and to ensure that there are sufficient resources to ensure activities for tourism development. They are also required to popularize tourism possibilities in their territories; ensure the preservation of the tourism objects and the opportunities of using them for the tourism needs.

A reasonable tourism development encourages the preservation of the cultural heritage. The most significant part of the heritage used in the historic centres of towns is architectural heritage. The totality of the architectural heritage should be regarded as of considerable economic value, used not only by its proprietor or other direct users, but also by tourists. This approach is equally important for Riga and other historic centres (Fig. 10.8).

A successful tourism policy gives a town an economical benefit through revenue derived from taxes. It is important to use a part of the attained financial resources for the preservation of the cultural heritage – in such a way as to qualitatively improve the tourism product.

It is important for the visitors to the historic centre to feel the atmosphere of the town, therefore not only are separate monuments important, but everything that creates a united image of the historic centre – buildings

Fig. 10.8 The reconstruction of Ventspils Castle.

with façade details along with their equipment in the streets and places, historical parks, monumental monuments, etc.

In Latvia, the Advisory Tourism Council of Latvia provides co-operation among the municipalities and administrative institutions of tourism, cultural heritage and environmental protection.

A balanced tourism development should be considered carefully to alter the attitude of the visitors who more and more support the idea of the preservation of cultural and natural heritage and ecology (Fig. 10.9).

These are the objectives of tourism policy for historic centres. As yet much work to integrate tourism and heritage management has still to take place. In the context of the proposed 'Preservation and Development Plan' this will be an important consideration for Riga.

Conclusions

Historic town centres are not fixed – they are under continual development. Each period of time leaves its traces on a town along with its qualitative supplement – layers. The claims on the preservation of cultural heritage may not be dogmatic: they have to allow the development that would preserve qualities. However, even modern development cannot be based on the destruction of the heritage of the past. In the proposed new plan for the historic centre of Riga, the following aspects will have to be considered:

- what should be preserved as a value;
- what should be eliminated as worthless;
- what should be transformed as unsuitable;
- what should be created anew as necessary.

Fig. 10.9 The timber housing of Jūrmala.

A gap between the historical and modern architecture is not admissible. Good modern architectures will serve as future monuments. The creation of a qualitative space is an important dialogue among all parties interested.

References

Historic Centre of Riga: Brochure prepared by the Nordic World Heritage Office (Oslo), Riksantikvaren (Direktoratet for Kulturminnevern, Oslo), and the State Inspection for Heritage Protection (Riga) in 1998.

Conservation and Development of Historic Centre of Riga 1995–2001: Information prepared by the State Inspection for Heritage Protection (Riga) in 1999.

Anthony Pace and Nathaniel Cutajar

Historic centre management in Malta

Introduction

The factor that mostly distinguishes the Maltese islands from many other countries is geographic size and population density. The Maltese archipelago consists of a number of very small islands of which three support a total population of about 378,518 (COS, 1998). With a total land surface area of about 122 square miles (246 square kilometres), Malta's population density amounts to about 3,100 persons per square mile. The larger part of the archipelago's population resides on the island of Malta, which is also the location for the major part of the industrial activity that takes place in the country. This socio-geographic context has necessitated political, legal and institutional frameworks that are very much particular to the realities of the Maltese islands. In the context of conservation policies, the dimension of legislative, policy and institutional frameworks are very tied to the scale of Malta's everyday economic and social activities.

In essence, issues related to heritage management and conservation have to be understood first and foremost within the context of small island states.

Since the 1950s, the economic history of Malta has been one of restructuring and diversification. This has been especially necessary due to the loss of British military spending after the Suez Crises (1956). Islands such as Malta and Hong Kong, as well as outposts such as Gibraltar, had for centuries served as strategic military and commercial bases. The long-term dependence on military investment of such bases often made economic diversification very difficult to achieve (Lowenthal, 1957). Paradoxically, the main part of the military investment went into the development of a formidable infrastructure of fortifications (walled cities, barracks, towers and related administrative buildings), that is now considered to be mostly of great value and an intrinsic part of Malta's cultural heritage.

Island economies suffer from a number of serious weaknesses that are not normally encountered in continental state economies (Brigulio, 1994). Islands suffer mostly from their smallness. Their geographic space is limited and finite. Often, access to islands is highly regulated through frontier checks or conditioned by distance and geographic isolation. Economic activity is often very small in terms of Gross National Product (GNP). Survival and development depend to a large extent on importation. Having small domestic markets, island economies have to depend on manufacturing industries that have to compete internationally. In effect, the fiscal possibilities for supporting conservation policies and interventions are in no manner proportional to the real needs of protection and preservation of the cultural heritage.

It is within the context of the disproportionate relationship between limited fiscal possibilities and actual conservation needs, that one has to consider the manner in which Malta's heritage protection institutions and operational frameworks developed.

Malta's cultural heritage sector is regulated by a small number of institutions and according to a series of legal instruments, some of which date back to 1925. The regulation of the Maltese cultural heritage was established on a proper footing during the first three decades of the twentieth century. For the archipelago this was a period of innovation and archaeological discoveries. In the space of a few decades a number of major archaeological sites were identified and excavated. In 1903, the archipelago's government established the 'Museum Department' to display government-owned museum collections and also to serve as the central government agency that had responsibilities over antiquities. To this day, the renamed 'Museums Department' (the plural denoting a number of 'off-shoot' museums that have developed since 1903) still has this twofold responsibility over collection management and display, and the management of the country's antiquities (Pace, 1998).

For about seven years, the Museum Department operated without any form of heritage protection legislation. In 1910, a Preservation of Antiquities Ordinance (*Malta Government Gazette*, 1910) was enacted. The Ordinance provided a simple framework for the protection of antiquities. It was inspired in the main part by the Italian legislation, which had been enacted in 1909. Following amendments and improvements in 1922 and 1923, a final Antiquities Protection Act was enacted in 1925 (Malta Government, 1925; Buhagiar, 1985). The 1925 Act provided for the establishment of an Antiquities Committee which, before being disbanded in 1992, assessed and advised government on the protection of heritage assets. It is the Antiquities Protection Act that to this day regulates some of the more important functions and procedures that concern the protection of cultural heritage.

Another 66 years passed before two new legal instruments, having some bearing on heritage issues, were enacted. The first of these acts was the Environment Act of 1991 (Malta Government, 1991). This law served to focus on environment protection, with provisions for the safeguarding of cultural heritage. In respect to cultural heritage, however, the Environment Protection Act did not suffice to provide a comprehensive framework that would have enabled broader heritage management structures to develop. Moreover, the Act was not considered to provide adequate cover for a series of important provisions that were enshrined in the Antiquities Protection Act of 1925. It is for this reason that the older act is still in force.

A year later, the Development Planning Act (1992) was enacted to regulate and establish modern planning procedures. The Act established a central Planning Authority as an autonomous agency to regulate development. Most of the planning principles and procedures were imported from British legislation and practice. Although greatly influenced by British practice, Malta's planning framework has paradoxically not adopted the critical presence and contribution of strong independent heritage institutions, such as English Heritage (Pickard, 1998). Such imbalances have served only to prolong an uneven development of Malta's heritage sector.

One recurrent tendency throughout Malta's history of institutional development is the reliance on centralized bodies. Although this trend was broken in 1991, following the overnight creation of local councils governing just under 70 localities, certain functions such as planning procedures, rehabilitation programmes and the regulation of heritage have remained within the immediate remit of central government. Thus responsibilities for conservation issues related to planning matters now reside in a centralized Planning Authority. Similarly, the Museums Department is the central agency that deals with heritage issues on a national basis. Rehabilitation programmes are again managed at a centralized level, but are directed towards localized areas that exclusively comprise Malta's walled cities.

For reasons outlined in this brief introductory overview, this contribution will be dealing mostly with national issues and practices for the management of historic centres rather than localized examples. Given Malta's circumstances, this approach may prove to be more beneficial even if in stark contrast to other chapters that deal with significantly different and diverse scales of operation.

The policy and planning framework

Recording the physical condition of the historic environment

Systematic detailed recording of architectural heritage is a relatively recent preoccupation in Malta. During the nineteenth and early twentieth century recording activities were heavily reliant on a monument-based philosophy and were particularly concerned with the more formal aspects of architectural value. Intervention was mainly limited to the Islands' principle high status buildings, with early conservation events being concentrated on such major historical landmarks as the Cathedral of St John's in Valletta or the Inquisitor's Palace in Vittoriosa.

Such early conservators, and later imitators, conceived their role as being primarily concerned with the recovery of important architectural works of art from a hundred years and more of colonial 'debasement'. The additions carried out by British colonial administrators to earlier Baroque and

Renaissance buildings were simply not considered as significant architecturally, and were normally wiped out without record. Conservation was perceived therefore as a 'restitution' of debased historical façades and interiors, through the removal of any intrusive built fabric, the uncovering of mouldings and frescos from layers of later plaster and through the profuse replacement of damaged or completely missing architectural parts. Budgets and mentalities were not really up to the task of inspecting and conserving the integrity or the authenticity of the architectural fabric *per se*. The 'archaeological' approach to the recording of buildings, respecting it as an organic product of different past generations was not practised.

The monument-philosophy – still popular with certain sections of Maltese society, as with the voluntary heritage groups – achieved some notable successes, uncovering and conserving for the Maltese public key examples of high architectural art, including medieval chapels, small fortified complexes and some important palace complexes. Yet on a broader scale, the monument-based philosophy of architectural heritage management suffers from a number of renowned shortcomings such as:

- detracting from the authentic multi-layered value of the architectural heritage, by concentrating on a perceived 'high moment' of the building's history to the detriment of all other aspects; and
- limiting the range of the heritage manager's activities to a few high status structures, thereby excluding from their action the majority of built heritage and the landscape context in which they are located.

Over the last 20 years, a growing concern with ensuring a more effective public management for the Islands' land resources has resulted in some radical re-thinking on the role of the architectural heritage in the modern urban landscape. These developments have stimulated a parallel concern with the quality of the available cultural heritage database and on the related record-taking systems. This trend has given rise to a number of different initiatives in the data-collection sector, and has also put in doubt the old monument-based approach.

Ever since the 1920s the main heritage database in Malta consisted of the Antiquities List, a legally binding document that purported to list all known archaeological and architectural monuments worthy of preservation – in emulation of the British Schedule of Protected Monuments. This important record of the Islands' architectural heritage, compiled largely in the 1950s, however, never went beyond the essential task of materially listing the buildings of special value. The photographic recording of the same structures and sites, initiated in the 1960s, similarly tended to take the short-cut expedient of mainly recording the façades.

In 1988, the Maltese government enacted the Building Permits (Temporary Provisions) Act in an effort to introduce an immediate control mechanism on development. The Act defined the boundaries of areas that were deemed to be suitable for development. The Act was designed primarily to discourage urban expansion, which had reached dramatic levels since the end of World War II. Current estimates project a short-term population increase of about 400,000 by 2020, giving an overall population density of 1,246 persons per km^2 and an urban sprawl that has engulfed nearly one-fifth of the country's land surface (Schembri, 2000). Containing urban sprawl meant that urban centres, including historic areas and traditional village cores, experienced a sudden surge of pressure as land suitable for development was immediately sought within them. The 1988 Act provided for the immediate designation of urban cores as well as areas that were suitable for development. Every city, town and village was simultaneously subjected to this new regime. While the effect was positive in its immediate curbing of development outside permitted building schemes, new factors impacted on the historic fabric of centres. Under threat of demolition or alteration, historic buildings experienced greater pressure as more and more households sought inhabitable houses.

The character of Malta's historic centres is dominated by urban development that followed the arrival of the Order of St John in the archipelago in 1530. In very broad terms, this phenomenon had a twofold effect. The first saw the transformation of the ancient prehistoric and ancient landscapes, parts of which had survived into medieval times, with the emergence of the new cities around the harbour area (the Cottonera walled cities), Valletta and Floriana (Pace and Cutajar, 1999) (Figs 11.1, 11.2 and 11.3). This Baroque infrastructure is rich enough in artistic and aesthetic elements to attract World Heritage attention as happened with the inscription of Valletta as a UNESCO World Heritage City, and the current tentative listing of the harbour area fortification ranges. The second transformation encompassed changes in the rural medieval landscape.

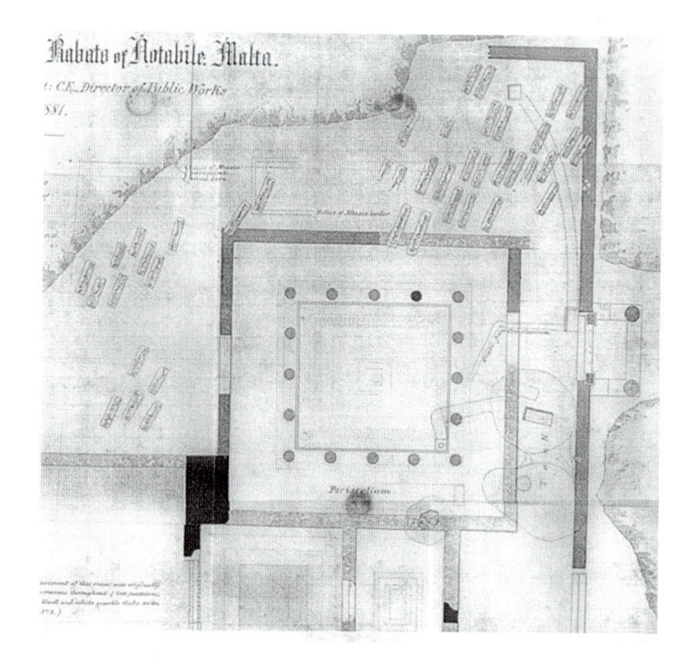

Fig. 11.1 The ancient town in Malta – a late nineteenth-century drawing showing an early Roman town house in Rabat with its colonnaded atrium and mosaics. The excavation of this site and its transformation into a museum display in the 1880s is one of the earliest instances of state intervention to preserve an aspect of Malta's urban heritage.

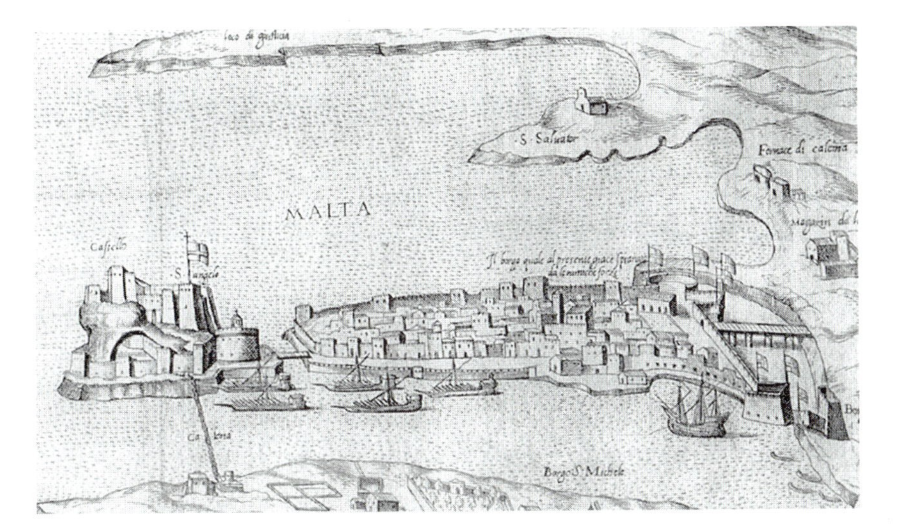

Fig. 11.2 A mid-sixteenth-century print showing the township of Birgu, Malta's principal maritime outlet throughout the Late Middle Ages. Typical of medieval urban form, the town is heavily overshadowed by the towering mound of Fort St Angelo.

Fig. 11.3 The modern town in Malta – an early seventeenth-century print showing the newly constructed city of Valletta with its powerful ring of fortifications and its orderly grid-iron street system. The city was much more than a mere administrative capital for the Order of St John, it also functioned as a potent icon of Renaissance ideologies.

Before 1530, a significant part of the archipelago's population was spread away from the established urban centres in small hamlets and settlements. The economic changes brought about by the Order of St John led to the gradual but steady abandonment of outlying villages and the consolidation of new areas most of which developed around the more convenient medieval villages (Pace and Cutajar, 1999).

The 1998 Building Permits (Temporary Provisions) Act brought all of this historic fabric into sharp focus. At the same time the establishment of the Planning Directorate, its successor the Planning Authority and the Publication of Malta's first Structure Plan (1992) generated well over a decade of data. Some stark contrasts were revealed, such as the high level of unoccupied houses. It is now estimated that the demand for housing throughout Malta will reach the 40,000 mark while at the same time an estimated 35,000 houses lie unoccupied.

The Development Planning Act and the Policy guidelines set in the Structure Plan led to the creation of a list of scheduled property and the establishment of Urban Conservation Areas. Whilst the schedule defined the graded status of listed properties in terms of cultural, historical and architectural value, the Urban Conservation Area provided a tool for a blanket restriction on development in designated protected areas. Throughout Malta and Gozo a three-level (1, 2 and 3) grading system classifies properties according to guidelines that have been established in the Structure Plan, itself a requisite under Maltese law. To date, about 900 cultural heritage assets have been scheduled; 70 per cent of these are historic buildings, mostly located within Urban Conservation Areas. The Planning Authority utilized technical parameters that are compliant with the guidelines contained in the Council of Europe's Recommendation No. R (95) 3 relating to the compilation of heritage databases and catalogues (Council of Europe, 1995).

Although effective as a development in the history monument protection, the scheduling process now requires a more qualitative examination of properties. To date, inspections and surveys of interiors are rarely carried out at the scheduling stage. Such inspections normally follow site visits by the Planning Authority's Heritage Advisory Committee as part of the planning process. Although important, this approach is somewhat retrospective in effect and does not allow enough manoeuvring space for planners and heritage managers to project long-term conservation strategies and possible usage. Furthermore, the lack of inspection does not enable a more objective audit of the importance of individual properties within a broader historical context.

In effect, the heritage inventory of the Planning Authority is not conceptually very different from the old monument-based listings. The new database primarily attempts to determine which built properties or urban spaces can aspire to an extended heritage status and identifies the selected properties by means of exact map references and of a fuller photographic record. In this sense the Planning Authority is merely using a more systematic version of the now defunct Antiquities List.

More complex approaches to data-capture and record-taking of the architectural heritage can be encountered in those public entities that are more directly charged with operational responsibilities. Particularly active in this respect are a number of regional/urban rehabilitation programmes and the Restoration Unit of the Works Division. This last entity in particular has been especially active in promoting the use of detailed mapping techniques – such as photogrammatic images transferred onto Geographical Information

System (GIS) mapping programs – as an essential prerequisite for the development and management of conservation projects.

An interesting development is the growing trend for conservation architects and archaeologists to work in tandem. Such collaboration helps create a better understanding of a building's history through the application of stratigraphic principles to cultural fabric as well as to the buried earth strata. This approach has allowed, for example, the identification and recording in some Valletta buildings of primal Renaissance architectural fabric, thereby distinguishing it from extensive later redevelopment in the Baroque and/or Colonial periods. The study made of the ground floor architectural fabric within the same National Museum of Archaeology (the former sixteenth-century Auberge de Provence) is a particularly clear example of this trend to apply stratigraphic principles to decipher a building's architectural history.

The preservation and enhancement of central areas, urban complexes and townscape and related urban design issues

The designation of Urban Conservation Areas as a blanket protection measure is effective at a macro-level. In reality the different conservation areas possess diverse qualities and therefore require different solutions to the application of conservation principles. In practice, Malta has six specifically designated Urban Conservation Areas apart from all the existing village cores. These are defined by policy UCO 1 of the Structure Plan (1992) as follows:

1 Valletta and Floriana;
2 the Three Cities: Vittoriosa, Senglea (Fig. 11.4) and Cospicua;
3 Mdina;
4 the Cittadella and its environs (Gozo) (Fig. 11.5);
5 the central area of Sliema;
6 the central area of Hamrun;
7 Village core areas.

At the macro-level, it can be said that these areas share a series of common problems. The pressures that were unleashed by the very designation of conservation areas and the definition of their boundaries has necessitated an evolutionary approach to development planning. As defined in the Structure Plan and Policy (1992) and Design Guidance 3 (Planning Authority, 1997), this evolutionary process should:

> aim to provide the right balance between the static system of restraints, that ensures the conservation of the inner urban fabric, and the dynamic system of stimuli that encourages sensitive and compatible urban renewal in line with the needs of modern society. The main goal of any urban conservation strategy should be to ensure that the evolution of traditional settlements is compatible with maintenance of the historical continuum. Evolutionary changes such as the re-use and

Fig. 11.4 The urban fabric of the Grand Harbour area due to World War II aerial bombing – the township of Senglea shown in the forefront.

Fig. 11.5 The Gozo Cittadella – the ancient and medieval administrative centre of the Island of Gozo and site of repeated attempts at urban rehabilitation by the Maltese government.

adaptation of historical buildings, the provision of social and community facilities, and the betterment of centuries old festivals, are not only desirable but necessary to sustain livability within the village and town context.

In essence, the policy guideline allows development in the form of re-adaptation and modern use of historic fabric as long as this does not impact negatively on the historical context of the area.

More specific problems provide a more tangible appreciation of the difficulties that are so often encountered in maintaining the right balance with conservation areas. The demolition of houses was by far the biggest impact. This phenomenon has been restricted as stronger conservation policies were established following the designation of the Urban Conservation area in 1988. Less difficult problems, but equally significant because of their collective negative impact, concern alterations to building height, façades or to architectural fixtures such as wooden balconies. The abandonment of traditional materials for the rendering of shop fronts and building façades followed. This was mainly as a result of a decline in design and aesthetic values, as well as the widespread introduction of non-traditional materials such as aluminium (instead of wood) and the use of steel, non-traditional rendering of stone, alien colour schemes and other elements.

Nearly all of Malta's historic centres and village cores experienced dramatic fluctuations in terms of population abandonment. During the 1970s and the first half of the 1980s, government policy favoured extensive, affordable housing schemes to counter growing social pressure. This policy encouraged urban expansion across tracts of landscape that were rendered affordable. An immediate result was the abandonment of historic centres. This trend is being reversed in varying degrees in some areas but not others, and this is as a result of the rise in real estate marketing of old houses. This commercial market has been made possible mostly as a result of the extensive alterations that are allowed in many of these historic properties. Being mostly of Grade 2 listing standard, the development of these properties often involves extensive alterations, additions and the creation of new spaces that are carried out without any notion of conservation standards and issues of authenticity. In most of these cases, the role of the Planning Authority is reduced to a retrospective consideration of development applications. These applications normally require inspections by the Heritage Advisory Committee. An immediate benefit of this real estate phenomenon has been a degree of re-population in historic centres.

Other historic centres suffer from different parameters. Valletta, for example, experienced a steady socio-cultural decline as a result of such factors as transformed entertainment markets, the domination of business and commerce in the central area of the city and the decline of traditional commercial areas. As the country's centre for government, commercial and banking activity, the city experiences a daily influx of workforces, a business and commercial community as well as a daily flow of vehicles. While extremely vibrant during the day, Valletta becomes a dead city after office hours. At the same time, however, the city also possesses residential areas that follow particular community activities in sectors that are often considered to be peripheral to the business district. Valletta's suburb, Floriana, is also enclosed within fortification walls. Like Valletta, Floriana follows an early grid system layout. Although marked by a significant number of monuments and Baroque buildings, Floriana suffers mainly from being a thoroughfare leading into the capital and is considered to be secondary to Valletta. The suburb has experienced depopulation, with several houses being abandoned, and a rise in office premises.

The problems experienced in the Three Cities area, across the Grand Harbour from Valletta and Floriana, are often perceived as being more acute. The area boasts an agglomeration of small towns that are tucked away behind the Baroque fortification ranges known as the Cottonera Lines. The origins of these fortifications go back to medieval times. The area has experienced much degradation and social problems. Severely damaged during World War II, the Cottonera was subjected to extensive construction that very often did not respect the Baroque character of the cities. The post-war reconstruction was directed towards the provision of cheap affordable housing. But this was achieved at the expense of the historical environments that were severely compromised. To complicate matters, the Cottonera area has historically been the location of Malta's docklands. This has meant that, in reality, the area had to receive within its historic environs a vast array of problems that are normally related to heavy industry.

Urban Conservation Areas and Structure Plan policies on their own are no answer to the pressing needs of residents and commercial concerns to re-develop historic architectural properties. As highlighted earlier, owners of urban properties can frequently argue that lack of land on the Island forces them to seek available space even within historical urban cores. This situation of direct confrontation between economic/social needs and conservation is often resolved by recourse to such debatable measures as façadism or by resorting to a mock-antique or 'traditional' architectural medium of construction to mask the wholesale erasure of old fabrics and of their related spaces.

Policy and practice are therefore not in perfect harmony in Malta when it comes to architectural conservation and heritage management. In fact a great divide can be traced in society's approach to land-development issues, particularly in such sensitive areas as historical urban cores. On one hand, one detects a strong pro-conservation sentiment which argues for a full respect of existing fabrics and of attendant urban spaces. On the other hand, one is confronted by a well-entrenched pro-development lobby which pushes for radical re-modelling of existing fabrics and land resources as the only means of sustaining the local cycles of economic growth and rising standards of living and consumption.

This dualistic situation is a reflection of the severe limitation on available land resources which affects every form of economic and political undertaking. A solution to this impasse is the full adoption and execution of integrated conservation approaches through specific action plans.

Management and regeneration action

The situation prevailing in Malta is one in which management and regeneration activities remain very much dependent on central government. The most significant departure from this norm is regeneration that follows market forces. Broadly speaking the regeneration of historical town cores can be described along three main categories:

1 Central government supported action;
2 Local government supported action;
3 market-driven regeneration.

Central government supported action
This includes a wide range of management situations, which collectively account for a considerable share of the government's annual budgetary allocations. In most cases, these initiatives are not co-ordinated and their execution depends on the annual business objectives of the different state departments. The principal forms of government-driven rehabilitation initiatives now fall into two main groups, those that are carried out by specifically established rehabilitation committees and those activities that are carried out by different government departments.

The rehabilitation committees have a specific organizational structure that distinguishes them from other state bodies. As organizations they are in the main autonomous, but operate within the Ministry for the Environment. Originally applied to Valletta, the idea of a rehabilitation committee has now been applied to Floriana, the Cottonera Area as well as to the ancient and medieval centres of Mdina (Fig. 11.6) and Rabat.

As their name implies, the aim of these committees is to manage rehabilitation programmes of the more important historic cities of Malta. The

Fig. 11.6 Aerial view of Mdina – Malta's ancient capital. This town core is an excellent example of the great diversity that characterizes Malta's urban heritage – from the deep lying Bronze Age strata to the splendid eighteenth-century Baroque cathedral.

experience of the Valletta rehabilitation committee has shown that the work of this organization has mostly focused on the restoration of particular monuments and its has been rather ineffective in applying an integrated conservation framework for the capital. The organization of the committees brings together a number of specialists as well as local government officials. The committees are, however, established by central government and are therefore independent of local authorities.

Apart from the work of the rehabilitation committees, central government also has a direct input as a result of departmental services. Agencies such as the Housing Authority, the Lands Department, the Works Division, Museums Department, the Malta Tourism Authority and others conduct several projects as part of their service requirements. Like the rehabilitation committees, initiatives undertaken by government agencies are not co-ordinated sufficiently to ensure an integrated approach.

Beyond government supported agencies, other initiatives that are currently being developed involve direct state encouragement of large-scale land redevelopment schemes by private or semi-private consortia on strictly commercial bases. Currently, one project that is gaining momentum is the Birgu Water Front project that involves the redevelopment of a substantial part of the Three Cities area. Although widely supported, the idea of 'smart partnerships' with private investors in the development sector is still at its infancy and the real long-term effects on the socio-cultural structure of communities living within historic centres cannot at this stage be evaluated.

Local government supported action

Established only during the 1990s, Malta's 67 local authorities are steadily being given authority over several initiatives. The remit of local government to manage certain aspects of streetscape design and to ensure that the local environments are maintained on a regular ongoing basis has become the key to launching heritage management initiatives in both urban and rural contexts. The particular strength of local governments is their ability to tap directly into local human resources, even on a voluntary basis, to draw attention and activity towards local heritage sites and spaces, not only through direct conservation measures but also by providing amenity facilities. Their particular weakness currently lies in the paucity of their budgetary allocations, and the insufficiency of their legal remits and management set-ups to handle heritage issues on an ongoing basis. In many ways the effectiveness of local councils lies primarily in the concern to manage the resources and assets that lie within local authority geo-juridical boundaries. In this context, heritage management concerns are rapidly becoming issues that define a sense of identity. Awareness and eagerness on the part of local authorities to protect cultural heritage assets are gaining more and more ground.

Market-driven regeneration

Market-driven regeneration is particularly carried out by small investors, typically small land or property owners or occupiers, investing in the upkeep or redevelopment of their own properties for residential or commercial purposes.

It is important to note that small-scale property owners are collectively in charge of the single largest share of the country's architectural heritage and urban spaces. Thus, for example, the impressive array of ecclesiastical properties on the islands, some dating back to the Middle Ages but most representing remarkable treasure houses of Maltese Baroque tradition, are held and owned entirely by small parochial or otherwise religious organizations. Similarly, the large stock of available historic housing – ranging from the small palatial complex with attendant gardens to the lower bourgeois landscapes of colonial suburbia – is managed and owned largely by the families who inhabit them.

It is in this sector of domestic real estate that a new vibrant form of regeneration is taking place. During the 1990s, the market for 'houses of character' and town houses has expanded rapidly. The market is, however, not regulated to conform to conservation requirements. Planning Authority policies allow the redevelopment of historic buildings, mostly Grade 2 properties, in the context of acceptable conversions. House conversions are by nature fluid and not tied very strictly to conservation principles expressed in international charters. This situation is a delicate one in that house conversion cannot be halted without compromising current trends in the re-population of historic centres. The critical problem is absence of any form of detailed inventory of house interiors that would help policy decisions in the case of development applications. As stated earlier, such applications often require site inspections by the Planning Authority's Heritage Advisory Committee. Although the process of considering an application is regulated by a framework of policies controlling procedures, property grading, the compilation of methods statements and other requirements (Planning Authority, 1997) there still remains the major issue of the individuality of every development application involving house conversion. While the real estate market in character houses is experiencing a boom and encouraging a certain type of regeneration, the sector remains problematic and certainly a dilemma for decision-makers.

Funding, society, environment and tourism

The separation of the roles of central government, local government and of the private sector has become a constant characteristic of Maltese heritage management operations in recent years. The funding mechanisms, social effects and economic success of each situation is likewise very different for each of the above sectors. It is also probably fair to say that any structural bridging-over of these positions has not been registered in the last 10 years.

These differences are instructive indicators of which basic mechanisms are in operation in the Maltese heritage sector. The main problems that currently seem to be facing the conservation, study and appreciation of Malta's architectural and urbanistic heritage can in fact be largely ascribed to non-integrated approaches to heritage management. The following examples characterize this situation in greater detail.

Funding
Expenditure of public funds in Malta is tightly regulated through a system of 5-year financial plans and through yearly budgetary re-assessments. This stable system of public expenditure relies heavily on the use of project-management techniques, and makes only limited use of such open-ended financing mechanisms as grants-in-aid programmes.

Such as system signifies that government is traditionally more likely to fund a few large monument-based restoration projects than to tackle broader issues of urban re-generation. Thus, for example, the bulk of the work of the Restoration Unit is largely taken up with the direct design, funding and execution of a very diverse range of specific conservation interventions. Prioritization in such circumstances is normally dictated by the gravity of documented structural problems, by the availability of public funding and by the logistical capabilities of the same unit.

Such a purely reactive system of intervention is difficult to avoid in a project-based public management regime. This means that while many emergency conservation programmes are successfully carried out each year through direct government funding, hardly any long-term planning and urban regeneration can be carried out through the same medium. Ongoing interaction with local communities is similarly unlikely.

Alternative forms of public funding for the promotion of urban rehabilitation are rare. An important start was made in 1998 with the launching of a grant-in-aid programme for the Cottonera Urban Conservation Area to assist private house owners to repair their antique wooden balconies according to traditional joinery techniques – a costly alternative to the cheaper, maintenance free, modern aluminum balconies.

The case of the Cottonera balconies was funded and managed jointly by the Planning Authority, the relative local governments and by the Cottonera Rehabilitation Project. The good results obtained and the popular success of this pilot programme must set us thinking.

The success of this grant programme demonstrates just how great is the need for government to move away from its traditional project-based, monument-philosophy in public spending policy for conservation and to concentrate instead on alternative forms of financing private and local government initiatives. It also illustrates the importance of networking between government bodies and local political realities in order to provide a greater degree of financial help and conservation know-how to the persons most heavily involved with the well-being of the urban heritage, i.e. the inhabitants of these centres and the people who work in them.

Social impact of urban regeneration initiatives
The social dimension and impact of urban regeneration initiatives varies greatly depending on which management approach has been adopted for its implementation. Once more the roles of central government and of local councils/private initiatives come across as being radically different.

As already stated, central government tends to concentrate on the restoration of key monumental sites, while largely failing to take a holistic approach to the historical urban fabric. Cities like Valletta and the Cittadella have seen a fair amount of work on major public buildings – churches, palaces, government headquarters – or on the more picturesque elements of ancient street furniture such as public fountains and commemorative statuary.

The social relevance of much of this publicly funded work is sometimes difficult to substantiate and assess. Thus, for example, a heavily worked-on town like Valletta is still losing its local inhabitants at an alarming rate. While the monuments of Valletta are gradually recovering from many decades of inattention, its living community is slowly dwindling away for lack of an authentic regeneration programme of its urban fabric.

What is missing are restoration projects and funding programmes specifically aimed at sensitively upgrading historic housing blocks, heritage industrial plant and storage vaults. The modernizing of these utilitarian, but architecturally valuable structures falls under the remit of different public agencies, not normally associated or used to dealing with the heritage sector – such as the Housing Authority, Ports Authority or the Malta Development Corporation.

This split approach to historic urban regeneration carries with it some serious repercussions. The run-down conditions of much seventeenth- and eighteenth-century housing in the Grand Harbour area has, for example, occasioned repeated calls to demolish them on account of their 'slum-like' conditions. The physical demolition of these districts is furthermore accompanied by the removal of local communities and their replacement with a new intake of residents, beneficiaries of social housing schemes. The erosion of human resources thereby follows closely on the destruction of the original architectural dimension.

By contrast the regeneration of town centres when managed by local interests (such as by local government or by local businesses) carries with it quite different social and conservation repercussions. The end of such exercises are typically either the upgrading of local living environments or the smartening up of localities to improve their business outlook and attractiveness (particularly in view of tourism or shopping nodes). A good case in point is presented by the re-birth of commercial activity in the historic core of Victoria (Gozo) over the last 10 years, the result of combined efforts by both the local council and of multiple small-scale private investors.

The impact on the ancient urban plan and on the social fabric is in these market-driven regeneration programmes rather low key compared to the government-sponsored gutting of perceived slum districts. The main threat to local community survival is in this case occasioned by the inhabitants' desire to sell out so as to capitalize on rising property prices. A further concern in such circumstances is caused by the fast rate of redevelopment, the difficulty to control such works and the low standards of many interventions.

Environmental management

The critical role of local councils also emerges clearly in the current situation with respect to traffic pollution monitoring and control. Local councils have been the driving-force in introducing one-way traffic systems, traffic calming measures and in the regulation of parking space. These changes were at times contested by residents and by local business communities. However, the desirability of such measures to improve living conditions in urban cores is being widely accepted and at times even urgently expected. In this regard, the Planning Authority has established important design guidelines that address practical solutions to traffic calming and also the creation of pedestrian areas (Planning Authority, 1997). These guidelines have been applied with good measures of success throughout the Maltese islands.

On the other hand, no ongoing campaign for the measurement of traffic-induced pollution exists on the islands. A few case studies have been carried out by the Department for the Environment in association with a select group of more exposed local councils. The data so far collected have, however, not been used in conservation-related issues, except in very few particular circumstances – as in the measurements carried out within the National Museum of Archaeology purely to establish the extent of damage to the National Collection as a result of atmospheric pollution.

Tourism and heritage and local identity

The link between tourism and the heritage sector gives rise to interestingly similar conclusions to the above discussion on urban regeneration initiatives. In fact the strong involvement or otherwise of government in the organization of heritage-tourism programmes seems to have a direct bearing on the social identity-content of these manifestations.

The rise of tourism as a powerful economic factor in the Maltese economy can be traced back primarily to the early 1960s. Government was then heavily committed to building up a viable tourism industry, which included its official support to specific heritage management initiatives. The concentration of national museums and collections in Valletta and in Vittoriosa is a case in point. Similarly, a stream of public money has been spent since the 1960s in grandly conceived restoration projects for the Cittadella urban core in Gozo.

These initiatives proved to be in the main good economic investments insofar as they, together with the more famous megalithic sites, attract over a million visitors every year, particularly foreign tourists. These visitor figures represent an important source of revenue for the government, as the income generated from the entrance tickets is currently hovering around the 800,000 Maltese Liri mark – a figure that only takes account of entries registered in state-owned museums.

However, beyond these statistical successes, the social dimension of the government-sponsored tourist attraction is somewhat disappointing. The case of the Gozo Cittadella speaks volumes. The site of a remarkable system of Baroque town fortifications and other architectural monuments, the Cittadella has been developed into a concentrated node of museological activities, nowadays containing three important state collections, as well as a Church-run museum. Tourism on the island of Gozo is in fact obliged to direct its attention to such a collection of heritage sites.

In spite of this effort on the part of government to provide centralized heritage attractions, the main social and economic life on the island is, however, not sited at the Cittadella, and is to be found in the medieval suburb of Victoria, about 1 km downhill. The regeneration of this important example of Islamic-style urbanism owes very little to central government's dogged investment on the Cittadella, but is largely due to local investment, as well as to improvements carried out by local authorities. Indeed, the majority of visitors to the Cittadella completely miss the lower town, even though in historical terms these two centres operate as an urbanistic pendant.

An alternative to the officially organized tourist attractions are the sites organized at grass-roots level. Such sites and events can be very successful in drawing large international crowds to historic urban and village cores, often on pretexts that officially have very little to do with the tourist industry. This variety of tourist–local cultural interaction tends to carry greater popular involvement and is more related to the people's actual cultural identities.

A classic case of the grass-root involvement with tourism is the outdoor village fiestas that are organized in various towns and villages on the occasion of major religious festivals or on the feast day of community patron saints. These boisterous nocturnal events, involving fireworks, music playing by brass bands and some inebriation, attracts thousands of Maltese and foreign visitors to each event. These occasions are mainly focused around the main areas of historic centres, where squares, the parish church and prominent buildings play an important role. The economic impulse at the village level represented by this major influx of visitors has become an important factor in guaranteeing the survival of these ever more grandly conceived celebrations. On the other hand, the foreign visitor is able to acquaint his- or herself with the Maltese urban landscape and society at a moment when its identity is most accessibly and extrovertly expressed.

Finally, it can be argued that the cultural heritage dimension of identity is a composite phenomenon that brings together what very often appear to be incompatible values (Pace, 2000). Whether advocating a case in favour of the cultural and aesthetic values as opposed to the commercial attraction of heritage, there remains the notion of an identity based on the sum total of contemporary interests in the past. It is this notion, and the viable balancing of competing interests, that can allow a meaningful approach to sustainability.

Sustainability

Sustainable heritage management is an elusive term especially in the context of development that, more often than not, elicits ideas of physical transformation or destruction of the heritage.

In a manner of speaking, sustainability of cultural heritage may still be taken to mean a viable manner of maintaining relevancy in the light of competing pressures. Although competing values can be categorized in several systematic ways (Planning Authority, 1998; Pace, 2000), it has to be recognized that certain values are in the main part transitory and, therefore, relative. But in the main part, it would be very unwise to equate sustainability of conservation with the concept of sustainable development, to which the conceptual terminology is clearly linked. It is clear that 'sustainability' acquires different meanings in the two contexts. Whilst sustainable conservation looks primarily towards safeguarding the past, sustainable development looks towards present requirements and future possibilities. Sustainable development has been defined as that which 'meets the needs of the present without compromising the ability of future generations to meet their own needs' (World Commission on Environment and Development, 1987). While this definition is designed to describe the manner in which development is ideally carried out, such a term as 'sustainable conservation' would, on the other hand, assume that the protection of the cultural heritage is equally a viable goal.

The dilemma, in this regard, is the planning methodology that is applied in order that the cultural heritage retains its significance and values in the light of requirements that are needed for urban centres to remain viably habitable. In particular, issues of authenticity and the unique character of cultural heritage become important factors that require close attention, planning and supervision in order that disfigurement, dilapidation, substantial alterations, change of character or demolition of buildings, spaces or streets are prevented (Yang and Taniguchi, 1999).

In Malta, the Planning Authority has adopted an evolutionary process of regeneration that allows changes in parts of the historic fabric to take place within historic towns and cities (Design Guidance 3, The Planning Fact Book) (Planning Authority, 1997). This approach is dependent mostly on the consideration of individual planning applications within much broader national policy frameworks of planning. As enshrined within the Structure Plan, the Development Planning Act (1992) and the workings of the Planning Authority, this pragmatic system of evaluating development requirements and pressures has greatly enhanced development control during the past decade. Such positive achievements are relatively new and recent, especially after decades of urban sprawl. However, dealing with individual pressures and applications may detract from broader conservation goals, as defined in the framework of integrated conservation (Council of Europe, 1976). While the success of Malta's planning process is laudable, the positive advantages gained during the 1990s over previous decades should provide a good platform for considering the future needs of heritage within historic centres. Following the successful integration of development planning policies during the 1990s, the next major challenge for the future is certainly that of applying a concerted approach to implementing integrated conservation policy packages in development practices. It is through the establishment of heritage assets as essential components of resource management and development that such goals can be achieved to meaningful levels.

References

Brigulio, L. (1994): *The Economy*, in Frendo, H. and Friggieri, O., *Malta Culture And Identity*, Government Press, Valletta.

Buhagiar, K. (1985): *Evolution of Conservation Legislation in Malta*, The Sunday Times Building And Architecture Supplement, Valletta, Progress Press.

Central Office of Statistics (COS) (1999): *Official Communication outlining 1998 Demographic Statistics*, Government Press, Valletta.

Council of Europe (1976): *Resolution (76) 28 Concerning the Adaptation of Laws and Regulations to the Requirements of Integrated Conservation of the Architectural Heritage*, Strasbourg.

Council of Europe (1995): *Recommendation No. R (95) 3 of the Committee of Ministers to Member States on Co-ordinating Documentation Methods and Systems Related to Historic Buildings and Monuments of the Architectural Heritage,* Strasbourg.

Council of Europe (1998): *Compendium of Basic Texts of the Council of Europe in the Field of Cultural Heritage*, Strasbourg.

Lowenthal, D. (1957): 'The Population of Barbados', *Social and Economic Studies*, 6 (4), Jamaica, University College of the West Indies.

Malta Government (1910): *Ordinance VI*, Malta Government Gazette No 5270, Government Printing Press, Valletta.

Malta Government (1925): *Antiquities Protection Act*, Malta Government Gazette, Government Printing Press, Valletta.

Malta Government (1991): *Environment Protection Act*, Malta Government Gazette, Government Printing Press, Valletta.

Malta Government (1992): *The Development Planning Act*, Malta Government Gazette, Government Printing Press, Valletta.

Pace, A. (1998): *The National Museum of Archaeology Permanent Exhibition Project*, in *Vigilo* (October 1997 – January 1998), Din l-Art Helwa, Valletta.

Pace, A. (1999a): 'A common space: the Council of Europe, Malta and the European perspective on cultural heritage', in Saliba, G. (ed.) *A Council for All Seasons: 50th Anniversary of the Council of Europe*, Ministry of Foreign Affairs, Valletta.

Pace, A. (1999b): *New Directions in Malta's Heritage Management* (forthcoming publication of Calpe 1999 Conference proceedings, Gibraltar).

Pace, A. (2000): 'Cultural heritage values as expressions of identity: the Maltese context', in *The Maltese Islands on the Move: A Mosaic of Contributions Marking Malta's entry into the 21st Century*, Central Office of Statistics, Government Press, Valletta.

Pace, A. and Cutajar, N. (1999): 'Malta', in *Report on the Situation of Urban Archaeology in Europe*, Council of Europe, Strasbourg.

Pickard, R. (1998): *Specific Action Plan For Malta*, Cultural Heritage Committee, Permanent Legal Task Force Legal Assistance (Technical Co-operation and Consultancy Programme), Council of Europe, Strasbourg.

Planning Authority (1997): *The Planning Fact Book, Malta*, The Planning Authority, Malta.

Planning Authority (1998): *Sustainable Conservation, A New Approach*, The Planning Authority, Malta.

Schembri, J. (2000): 'The changing geography of population and settlement in the Maltese Islands', in Vella, C. (ed.) *The Maltese Islands on the Move, Valletta*, Central Office of Statistics, Malta Government.

Vella, C. (ed.) (2000): *The Maltese Islands on the Move*, Valletta, Central Office of Statistics, Malta Government.

World Commission on Environment and Development (1987): *Our Common Future*, Oxford University Press.
Yang, M. and Taniguchi, J. (1999): *Implementation of the World Heritage Convention – Seminar on the Development and Integrity of Historic Cities*, Nara, Japan.

Xerardo Estévez

Santiago de Compostela, Spain

Introduction

The city of Compostela was founded in the Middle Ages, springing up from a small prehistoric settlement that had scarcely been influenced by the Roman occupation. In the ninth century, the *inventio* or discovery of the presumed remains of the Apostle James (*el Apóstol Santiago*) gave rise to a Europe-wide religious, political and cultural phenomenon, namely the pilgrimage to Santiago de Compostela. After the Basilica of the Apostle was built, the town centre was gradually consolidated around it and the first ramparts were constructed (Fig. 12.1). In the twelfth century, under the episcopate of Diego Gelmírez, Compostela achieved city status. Due to its increased size, and thanks to the 'Road to Santiago' de Compostela, many foreign architects, mainly Frenchmen, settled in the city to plan major ecclesiastical buildings. The Gothic style was introduced in the thirteenth and fourteenth century by the mendicants, and around this time the city began expanding beyond the city walls. In the late fifteenth century the university was founded under the patronage of Alonso de Fonseca, thus initiating the Spanish Renaissance.

With the Reformation in the following century, European pilgrimages went into decline. However, at around the same time Spain began colonizing America. The cult of the Apostle James crossed the Atlantic to the new continent, as demonstrated by the fact that over 200 Latin American cities bear his name, e.g. Santiago de Chile, Santiago de Cuba, Santiago de León de Caracas, Santiago del Estero, Santiago de Querétaro, Santiago de los Caballeros. Throughout the sixteenth century the city continued its gradual growth, but the real architectural boom came in the seventeenth century, when the Baroque style completely transformed the urban scene. The ecclesiastical class, which still governed the city, promoted this transformation thanks to the enormous income received by the Church from *ex voto*

Fig. 12.1 View of the Cathedral and a part of the historical centre.

offerings to Saint James. Thanks to these funds the Church brought in a fresh wave of foreign, Castilian and Andalusian architects, who trained the first generation of local artists, creating new, specifically Galician art forms. The Baroque reform continued during the neo-classical period, when many of the medieval façades were demolished and the historical ensemble was finally configured as we know it today.

Subsequent alterations were less obvious. In the nineteenth century the interior decoration of the buildings was modified and the ramparts and many of the arcades flanking the streets were demolished. However, no major changes were made to the layout of the city, as can be seen by comparing the maps drawn by Laforet, Cánovas and de la Gándara in 1908 and by López Freire in 1796. This state of affairs subsisted until the middle of this century. In 1944 the city was declared an official historic and artistic ensemble, and in 1947 a plan was issued for further development. In 1966 a *General Plan* was approved and implemented through a number of subordinate sectoral plans. One positive aspect of this plan is that it provided for protected areas, but it had the drawback of cluttering up land available for building. In 1974 this plan was revised on the basis of regulatory development-oriented criteria, but unfortunately the text lacked any real enforceable provisions. This led to increasing speculation in the development area, encouraged by the population growth, the boom in the building sector and the considerable economic feedback from emigrants to central European countries. Between 1970 and 1991 the population increased from 71,000 to 106,000 as a result of the larger university intake and the enactment in 1982 of the law granting the city the Status of Capital of the Autonomous Community.

The policy and planning framework

The 1979 democratic elections marked the turning point for the introduction of a new urban development policy. In 1985 the law on the Spanish Historical Heritage was adopted. This text, which complies with the relevant European conventions, sets out the framework for protecting the cultural heritage and enshrines the urban development plan as an instrument for architectural, environmental, social and economic protection. At the end of 1985 UNESCO included Santiago de Compostela in the World Heritage List. In 1988 the revised General Urban Development Plan (*General Plan* – Fig. 12.2) was approved, expanding the area of the historic city to 170 ha, and in 1997 the Special Protection and Rehabilitation Plan (*Special Plan* – Fig. 12.3) for the historic city was also approved.

The General Urban Development Plan: aims and strategies

It might be useful to outline the *General Plan* at this point, given that it was used as the reference for subsequent planning work in the historic centre. The culmination of 12 years of urban development work, the plan had major contributions from the architectural sector and very close co-ordination between town planners and architects. It was drawn up by a team that remained constant over the 12 years, namely the Planning Office, with contributions from Josef Paul Kleihues in respect of the *Special Plan.*

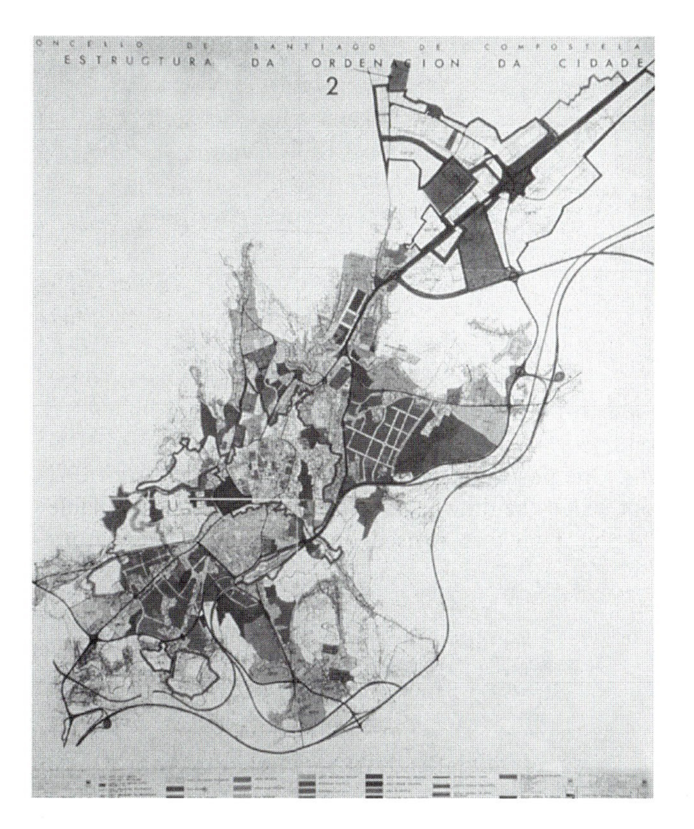

Fig. 12.2 The General Urban Development Plan.

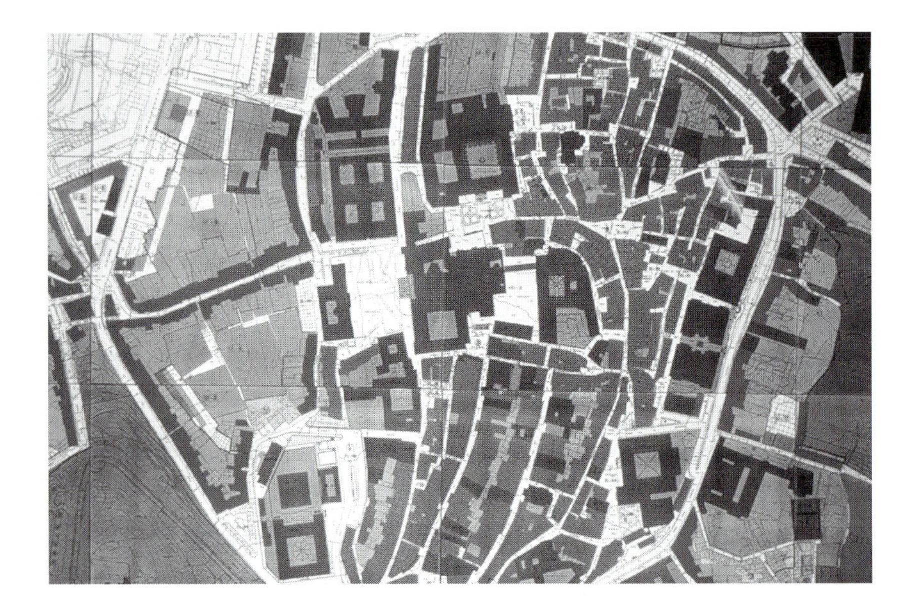

Fig. 12.3 The Special Plan for the historic city.

The aforementioned new challenges to Santiago de Compostela necessitated an immediate response. Every town or city is the result of innovative thinking, but implementation of any such project necessitates intimate knowledge of the area, *inter alia* from the historical angle. The transformation had to be effected according to a very precise schedule in order to meet the needs of a twenty-first-century city where history could provide for future impetus rather than stasis. The city had to be opened up and new economic channels generated within the university, the business sector and local government, implementing sustainable, high-quality policies and preventing social exclusion, etc., with new, though not necessarily original or ostentatious, types of architecture. The transformation had to be more obvious on the outskirts of the city but more rehabilitation-oriented in the centre, linking up the urban and rural environments, a process to which Santiago de Compostela lends itself quite naturally. The whole development approach also had to be based on the European spirit, which floods into the city along the Road to Santiago, using a multicultural language.

In physical terms, the development of Santiago de Compostela follows an extremely simple plan. The city centre runs north-west to south-east, and on either side are two areas that were singled out for special treatment. To the west is the enchanting area around the Cathedral constituting the historic centre, surrounded by a green belt along the banks of the river Sarela. This area, where the new university campuses are sited, is an architectural and environmental protection zone. To the east are the new housing and industrial estates, stretching along an 8-km north–south trunk road, which runs parallel to a non-toll-paying motorway taking traffic from the north to the south of Galicia. The installation of economic and social activities along this functional road system, which we shall describe in greater detail in the section on mobility, has greatly enhanced the area in socio-economic terms. It has also to some extent curbed the speculation that had characterized urban development in the 1960s and 1970s, when public investment was used to promote private projects.

In addition to these developmental criteria, general traffic and public transport has been properly organized and functional and productive capacity regulated and improved. Tertiary activities have been decentralized and alternatives are being sought for operations incompatible with the historic fabric of the centre. All such activities are relocated in other areas of the city.

The state of the historic city and the heritage

Up until the development boom in the 1960s, Santiago de Compostela had a similar profile to other small European towns in residential, functional and social terms. From this decade onwards the tertiary sector grew enormously in the city, prompting the traditional type of population to move to new areas, to the detriment of the architectural heritage and the urban environment. At the same time, the areas adjacent to the historic centre also began to deteriorate. The housing boom at the beginning of the 1980s did not help the situation: 16 per cent of dwellings in the historic city remained unoccupied and the percentage of elderly inhabitants topped 34 per cent, a situation that persisted until the adoption of the *General Plan*. In 1985, when Santiago de Compostela was included in the UNESCO World Heritage List, the revision of the plan was initiated.

For all that, the historic city has managed to preserve some exceptional assets to this day:

- it has a balanced combination of large-scale public, religious and civil edifices and residential buildings that boast different, indeed sometimes contradictory, architectural styles while still maintaining a basic unity;
- it adapts to its topography, perfectly linking up the ornamental monastic architectures and the compound walls with the 8–9 metre façades on the residential architecture;
- the architecture is constantly being renovated using traditional methods and avoiding any outdated heavy industry which has destroyed so many towns and cities; it has in fact managed to stabilize the historic image of the city as it was at the beginning of the twentieth century;
- it has simplified the use of quality materials: wood, stone, tiles and iron;
- the whole complex is unified by greenery, harmoniously blending town and countryside;
- without rejecting outside influences, the city maintains the specificity bestowed upon it by its inhabitants.

The Special Plan: objectives and regulations

Transformation (in the sense of reconstruction, replacement and extension rather than demolition) and conservation are the two dimensions of what is clearly a dialectical process, which currently offers a variety of theoretical options. One salient solution was defended by Rossi in the 1960s and 1970s as a response to, or rather a protest against, the excessive modernity of the modernist movement and the predominance of economic interests in the 1950s and 1960s which threatened the survival of the historic city. However, his theories subsequently achieved a kind of cult status, and virtual imitation of the buildings he constructed. The *Special Plan* for the historic city of Santiago, as drawn up by the Planning Office under Kleihues' supervision, was based on critical reconstruction. As clearly stated by Kleihues himself: 'modernity is a slice of the history of our culture and lives, which is why . . . we must be aware not only of a contradictory and complexly organised tradition, but of the modern protest against that tradition' (Kleihues, 1995). Such critical reconstruction is based on a threefold strategy:

1 analysing the city surveys, which give concrete expression to its spirit and culture by leaving the archaeological imprint of all the phases in Santiago's history;
2 analysing its stereometry, i.e. its geometry and construction activities; and lastly,
3 analysing the city's image, its physiognomy. This sometimes prompts action similar to that required in buildings, i.e. reinforcing part of the fabric or restructuring other parts in order to highlight a better profile or add an extra storey if needed, etc.

The *Special Plan* for the historic centre involves work over a total area of 170 ha. When the information survey was carried out for the development plan the area included 2,859 buildings, comprising 6,484 dwellings, of which 557 were vacant, and 1,883 commercial premises, of which 278 were unoccupied. It emerged from analysis of activities in these establishments that 40.6 per cent were involved in trade, 26 per cent in the hotel business, 13.5 per cent in services and offices and 19.9 per cent in other activities.

The report on the plan mentioned the following aims:

3.1 *Protecting the architectural heritage*. The catalogue lists 41 monuments of the utmost architectural and historic value; 68 exceptional buildings of great architectural and historic value; 296 buildings of particular interest; and 1,411 buildings of interest. The specifications for restoration, conservation, protection and rehabilitation cover all four categories, though the last one excludes buildings that do not fit in with their contexts, which facilitates a more flexible blend of modernity and identity. For all other buildings in the centre, detailed regulations have been devised in keeping with their structure, category, distribution, composition and physical make-up. In accordance with the principles of the ICOMOS Washington Charter, the original division into plots of land is preserved, as are the interior and exterior architectural features, and traditional materials are treated appropriately with new techniques in order to improve living conditions. The catalogue also lists areas that are differentiated according to historical, typological and architectural formation, and defines preferential uses of land and buildings in the zone.

3.2 *Rehabilitation of the central functions and future economic and institutional uses*. In accordance with the guidelines set out in the General Plan, the major industrial amenities are located outside the boundary of the historic centre. Inside the boundary, the institutional buildings are accorded special treatment, and the university is to increase in area from 19,500 m^2 to 47,500 m^2. In connection with cultural and tourist infrastructures, the aim is to promote the existing amenities and create new ones by either rehabilitating monuments or building new facilities, as in the case of the Galician Contemporary Art Centre.

3.3 *Improving residential uses*. Drawing on the traditional types of building and division of land, a general housing programme has been drawn up for the next 12 years. Fifty per cent of the cost is to be subsidized with public funds, covering the rehabilitation of 1,273 buildings (accounting for 50 per cent of housing stock), with some 2,407 earmarked dwellings. In the remaining spaces 503 new dwellings are to be built, and the corresponding amenities are to be provided or improved, doubling the existing public investment and increasing the green areas from 21.3 ha to 30 ha. A green belt is also to be installed around the historic centre, bringing the total green areas, combined with those on the university campuses not included in the General Plan, to 500,000 m^2.

3.4 *Developing pedestrian areas*. The growth model laid down by the General Plan facilitates continuous, dense and compact development, emphasizing pedestrian facilities and establishing pedestrianization policies, which we will go into in greater detail below.

Rehabilitation operations: some examples

With a view to implementing the various levels of action pinpointed during the drafting of the *Special Plan*, it was decided to bring in prominent architects from Galicia and abroad. These operatives had to have wide experience in similar operations and guarantee the incorporation of new contemporary contributions into the city's architectural heritage. Some examples of these operations are described below, highlighting the following aspects:

- the dialectical and simultaneous combination of general development and the specific architectural projects;
- the role played by architecture in rehabilitating the surrounding areas, both in urban and in socio-economic terms, and its positive effect on the environment;
- the role in repairing the occasional damage to the historic fabric caused by past speculation;
- specifically functional aspects as mechanisms linking up and assigning specialist roles to different parts of the city.

In order to repair the torn fabric and open up new prospects for the historic city, architecture and development have a specific role in 'stitching up' or restoring the damaged fabric by creating new spaces.

Remodelling Juan XXIII Avenue, 1996

In the 1960s asphalt made its first appearance in the historic city, bringing tourist traffic with it through the main point of access, which cut through a series of *rueiros*, or streets lined with traditional buildings radiating out from the higher parts of the town. The City Consortium sent out an invitation to tender relating to a project to solve this problem. Piñon and Viaplana secured the contract, and remodelled the existing road network by rehabilitating some of the old pathways through the town (Fig. 12.4). A three-tier service system was built up: at the lowest level was a depot for tourist buses, which had previously been allowed right into the historic centre; on the middle tier was the public reception area and a car park; and at the top was a 300-metre-long pedestrian boulevard with a fine glass and steel roof sloping downwards in order to direct attention towards the historic centre. This operation was complemented with a new sports centre for a nearby school.

Santiago de Compostela has always seen controversies whenever new buildings were to be superimposed on others. Throughout history, opinions have invariably diverged on new constructions. Nowadays, even though certain new buildings can have a monumental or exceptional character, most of those around the historic centre tend towards discretion. The question is whether ostentatious new items should be allowed to burst in upon the urban scene and, if so, on whose authority. The only possible answer is that the authority must be provided by planning and projects, although time, too, is an excellent ally in such matters. With time, opinions and viewpoints evolve, so that the public eventually come round to supporting, or at least tolerating, the work.

San Clemente Sports Centre, 1995

This project was aimed at building a new school facility on a site near the historic city, beside a historic monument, the *Colegio de San Clemente* (St

Fig. 12.4 Juan XXIII Avenue; in the background, the convent of San Francisco.

Clement's College). It also involved improving local amenities, with the construction of an underground car park, commercial premises and a small public square. Opinions had previously diverged concerning the use of the Colegio site: the great majority wanted the building to be converted to house the Regional Parliament, whilst the municipal council and the schools wanted it to continue as a school. But there was an overriding need for public educational amenities in the historic centre. It was agreed that these services should be improved, simultaneously helping to bring about policies for organizing road traffic and pedestrian circulation. Josef Paul Kleihues was therefore commissioned to build a sports centre and an assembly room, together with a 400-space car park.

This operation predictably prompted opposition from some specialists, who spread the idea that the new building 'would obscure the spires of the Cathedral'. It was even rumoured that UNESCO would be induced to remove Santiago from the World Heritage List. The subject was discussed extensively by the Spanish Academy of Fine Arts. The main promoters of the project explained how it fitted into the plan, with the result that it was unanimously accepted. The UNESCO Secretary General himself, who was visiting the city, was greatly in favour of the operation. In fact, the controversy was not, as had been claimed, about the height of the building or the suitability of the materials used (the design of the glazed gallery and the copper roof, which had never been used in the historic centre before), but about a matter of hierarchy: that of the Committee on the Historic and Artistic Heritage or that of the *Special Plan*.

Galician Contemporary Art Centre and the Bonaval Park, 1993

This was a highly committed venture to build a contemporary arts centre alongside the ancient Gothic and Baroque convent that currently houses the Galician Ethnographical Museum (Fig.12.5). The municipal council promoted the project and provided the land for the construction of the Galician Contemporary Art Centre, which the Galicia Regional Government commissioned from Álvaro Siza.

The new building put the finishing touches to a street which, like Juan XIII Avenue, had grown up within a particularly rich historico-cultural context. The building was V-shaped in order to incorporate a small inner courtyard and also to align with the existing street layout. Siza tailored his project to the *Special Plan*: he harmonized the roofs of the new buildings with the existing ones, thus enhancing the historic building and contrasting it with the rather bare modern façades. The white and timbered interior areas, covering three storeys, are unusually intimate for this type of construction. The approach was quite uncompromising: the modern style is, quite simply, modern.

The same architect also co-operated with Isabel Aguirre on other local projects conducted by the municipal council, amply but modestly demonstrating his great ability. The old convent grounds are divided into two sections, namely the decorative and vegetable garden, and the area that served as the municipal cemetery from 1837 to 1960. The funerary archi-

Fig. 12.5 Galician Contemporary Art Centre, next to the Museum of Galician People.

tecture, which is very important in Galicia, was carefully preserved in this part of the grounds. In the other section, partitioned off by a massive wall, the old compound walls are being restored and remodelled to form an elegant ascending sweep. In accordance with the guidelines of the *Special Plan*, the historic centre has thus been surrounded with parks so that it can be viewed from all angles, rehabilitating and enhancing all the various perspectives.

Moreover, the new high-quality architecture in the other municipal districts has improved the layout of the sprawling outskirts (this is a typical feature of Galician towns), increasing the value of property and attracting economic activities. It has also provided new services and increased the quality of life, with a view to attracting new residents.

Primary school, 1995

In an area located between the Northern and Southern Campuses, the Galicia Regional Government and the municipality adopted a joint plan to build a primary school complex once the banks of the river Sarela had been developed under the *Special Plan*.

School architecture often goes by the board in urban planning, but in this case Giorgio Grassi took particular care with the school building. The simple forms used for both the ground plan and the elevation, together with the excellent choice of site, have enhanced the whole architectural heritage of the surrounding district and prompted improvements in infrastructures, with additional private urban development.

The Galicia Auditorium, 1989

The so-called *Burgo de las Naciones* ('international village'), a residential operation planned in the 1960s to accommodate the mass influx of tourists, was transformed into the Northern Campus under the *General Plan*. A subsequent detailed survey defined the academic and cultural uses of the area. The nearby district of Vite was an example of the type of housing estate (in fact it was the first to be built in Spain, in the 1960s) with large neglected open areas, low-quality housing and major social problems. The urban development and architectural work which was subsequently carried out turned the estate into a high-quality area within a relatively short time-scale with much less risk of social conflict. This work embraced the faculties, the auditorium (Fig. 12.6) and halls of residence, and simultaneous improvements to local amenities (schools, sports facilities, health centre, socio-cultural centre and green areas) and housing rehabilitation.

The chosen site for what was the first large-scale operation under the urban renewal project was a low-lying area alongside the historic centre, where a music and plastic arts centre was constructed. The successful bidder for the contract, Julio Cano Lasso, designed a soberly configured building of stone, crystal and copper. The centre combines a rational volume in its central quadrangle and around the decorative pond, which magnificently reflects the prism housing the theatre. A few concessions were made to traditional Galician architecture such as the gallery along the façade overlooking the historic centre and a certain Baroque touch in the interior décor.

In Santiago de Compostela, the city and the university have formed an inseparable whole for five centuries now. This was the reason for a joint urban development project creating public areas to be used by both the townspeople and the university population. In view of the social importance and prestige of the university functions, it was decided to build the new Northern Campus, connected to the central and southern parts of the university by a pedestrian corridor running between the Pedroso hill and the historic centre. This western area is one of the most beautiful parts of the city in environmental terms, and efforts have been made here to protect this exceptional landscape.

Fig. 12.6 Auditorium of Galicia, surrounded by the 'Música en Compostela' park.

Vista Alegre University Park (under construction)
A former private property is the planned site for the construction of three small university buildings (Centre of Advanced Studies, Library and Academy of Music) and a simple teachers' residence. The original building is to be renovated as a European Centre, dovetailing with an area of traditional housing on the edge of the historic centre.

Arata Isozaki's project paid particular attention to the integration of a private garden into the ensemble of green areas surrounding the historic centre, providing a new perspective on the mystery of the monumental complex, these grand architectural flourishes superimposed on exceptionally beautiful creations of various different styles.

Management and regeneration action

The City Consortium as a grouping of different administrative departments

If we had to define the optimum therapy for the preservation and enhancement of the historic city, we could sum it up in three major strands: planning, the political will, and the requisite consensus between administrative departments and the various political tendencies represented in the municipal council. The political will in the municipal council is of overriding importance in order to be able to fully implement the requisite guidelines (this must be unflagging and meticulous in terms of dialogue with citizens to resolve their problems).

Thanks to the Jubilee Years of 1993 and 1999 and the inclusion of Santiago as one of the European Cities of Culture in the year 2000, a special programme has been conducted known as 'Compostela 93–99' aimed at transforming, but also conserving, restoring and rehabilitating the city's heritage. A special heritage body was set up for the purpose, the *Real Patronato de la Ciudad de Santiago* (Royal Patronage of the City of Santiago), under Crown auspices. This body is presided over by the King of Spain and comprises representatives of the three levels of government (local, regional and national), as well as of the university and the Church. Its basic aim is to channel all the general programmes into one overall strategy and for all three levels of government to implement this on an annual basis. The scheme covers both major 'investment', which is normally earmarked for development areas outside the city walls, and 'operations' inside the historic centre. The major investment effort was virtually completed in 1998 and has given the city the requisite amenities and infrastructures to fulfil the new roles assigned to it. The activities in the centre have been smaller in scale and more concerned with conservation, restoration and rehabilitation of buildings, and will continue over a longer period.

In order to implement these programmes, the 'Real Patronato' set up an executive body, the City Consortium, chaired by the Mayor and comprising nine members, three for each level of government represented. In the case of the municipal council, this meant ensuring that all its component political groups were represented. On the budgetary front, the state contributes 60 per cent, the regional authority 35 per cent and the municipality 5 per cent of overall funding. The body operates on the basis of the consensus principle, whereby decisions must be adopted unanimously. Broadly speaking, responsibility for proposals and suggested new projects goes to the local authority. Every year at budgeting time, the latter presents its proposals, most of which are based on the planning documents and are therefore definite necessities rather than mere showy extemporaneous ideas. Such solid proposals facilitate the search for consensus among the three government levels.

These operations have been successful because the project concept has been fully supported by all the parties involved (through the dialogue amongst different levels of government, the governing board, and a willingness for the government and opposition to co-operate), apart from the intrinsic potential of the historic city itself.

In 1998 the local authority presented a new programme entitled 'Compostela 2000–2004', complementing the previous 'Compostela 93–99' venture. The year 2004 was chosen because it will be the next Jubilee Year. The planned investment for the programme is mainly directed at the historic centre, including conservation, restoration and rehabilitation of monuments, traditional housing, infrastructures, and such movables as altarpieces, organs and archives, and ordinary expenditure on cultural and tourist activities.

The Rehabilitation Office: the 'bridging plan' and other programmes

One of the Consortium's most interesting initiatives was the setting up in 1994 of a subordinate Rehabilitation Office, comprising a small but competent and enthusiastic group of architects, economists, archaeologists and documentalists, who provide the real impetus behind the rehabilitation effort. The Office's working methods fully comply with the criteria the Amsterdam Declaration (Council of Europe, 1975) concerning the use of integrated approaches to the conservation of the built heritage. It prioritized action through direct, personal contact with owners and residents with a view to preserving the social fabric of the historic centre and creating the right conditions for perpetuating it.

While the work was proceeding under the Special Plan during its implementation phase, a 'bridging plan' was introduced to promote rehabilitation in the broader sense. One of the first challenges was to create or re-create transitional heritage conservation and maintenance techniques. Rehabilitation is not so much a question of materials or subsidies as of a prior interdisciplinary analysis, collecting data and assessing technical, economic and social problems with a view to providing an appropriate response to the problems of the historic centre. The operation in the historic centre was extremely complex, requiring painstaking and rather unglamorous work from the architect. The campaign to restore and preserve this heritage site began with work on the outside of the buildings: façades, drainpipes, timberwork, etc. The second phase involved successive 'bridging programmes', which were not only practical but also pedagogical, persuading the citizens of the benefits of rehabilitation and of the need to become involved in the assignment jointly conducted by the public authorities and the technicians. The public thus discovered that the operation concerned the whole personal and collective history lying beneath the successive strata of the buildings and surrounding area, with a concomitant attempt to re-establish harmony between architecture and interior design.

By setting up the Technical Office, the Consortium was launching an ambitious plan to harmonize the requisite maintenance of housing with respect for and conservation of the items characteristic of the various types of architecture. The Office provides individuals with the necessary technical and administrative assistance to estimate the scope of the work and the budgets and to choose the right company for the job. Thanks to the close contact between the technicians and the citizens, the vital preconditions for integrated conservation of the heritage can be reconciled with the residents' needs and means.

The Office undertook to organize the operational bodies involved in the process. The programme was implemented by some thirty officially approved enterprises, which accepted the budgets earmarked for each operation as a maximum. The enterprises also improved their qualifications through regular practical courses on techniques and materials and the experience gained from the programmed activities, under constant supervision from the Office. City architects are totally committed to the programme, working on a rota basis, whilst other young architects are completing their academic training and acquiring specialist knowledge through the *Aula de Rehabilitación* (Rehabilitation Workshop), and helping draw up projects and direct works.

All the projects are developed in the Technical Office, apart from those whose complexity places them in the 'major works' category (these are assigned by owners to architects of their own choosing). At all events, the Office plays a vital educational role, explaining directly to those concerned how and why the rehabilitation is being carried out. Furthermore, since the Office provides for 'one-stop shopping' (with the technicians from the various competent departments forming a single working committee), it enables owners and residents to easily overcome any obstacles that might arise, or even, if appropriate, to dissuade them from proceeding with any work at all. It not only issues the plans for the operation but also facilitates the procedure for obtaining municipal permits, the prior historical heritage report drawn up by a team of archaeologists, a list of qualified enterprises, information on financing, etc. This provides for a system based on rapid administrative formalities, complete technical assistance and direct personal contact with each of the operators involved. Last but not least, it is also based on lightweight intervention and flexible, 'job-by-job' rehabilitation planned in accordance with the guidelines of the Special Plan. Lightweight rehabilitation is seen as an optimum instrument for conserving historic cities, contrasting with the 'macro-operations' conducted under some urban renovation plans. The 'scattered' nature and number of the operations ensure proper diversity, unlike certain major projects with their unnecessary standardization of solutions and programmes. By the same token, investment in housing is affordable with or without loans, and the scheme is generally considered positive (Fig. 12.7).

Fig. 12.7 An example of interior intervention carried out by the Rehabilitation Office.

After an initial phase of implementing the housing programme, a parallel operation was launched on the commercial premises, improving their façades and installations and bringing them into line with security and health regulations. In a subsequent phase, the work was extended to restoring and conserving specific buildings: theatres, markets and churches. The figures set out in Table 12.1 and 12.2 summarize the overall results.

Table 12.1 Housing programme (November 1994–August 1999)

	Pesetas	*Euros*
Total works completed	1,832,318,174	11,011,527
Total non-subsidized works	1,212,371,275	7,285,884
Total subsidized works	619,946,900	3,725,643
Number of completed works	538	
Average amount	3,405,796	20,468
Average subsidy	1,152,318	6,925

Table 12.2 Commercial premises programme (January 1996–August 1999)

	Pesetas	*Euros*
Total works completed	169,192,156	1,016,788
Total non-subsidized works	123,203,319	740,403
Total subsidized works	45,988,836	276,375
Number of completed works	51	
Average amount	3,317,493	19,936
Average subsidy	901,742	5,420

Most of the permits issued were applied for privately, without using subsidies from the Rehabilitation Office, and in the 2 years since the approval of the Special Plan culminated in the following completed projects:

- 90 buildings;
- 323 dwellings;
- 435 commercial premises;
- 8 hotels.

These figures show the importance of the activity in the historic centre, in urban development and economic terms.

A number of conclusions have been drawn from the results of the rehabilitation action, showing that new technologies, if coherently implemented, can satisfactorily solve the problems of historical buildings and the conservation of traditional structures, without altering the essence of the heritage. Some examples have been: the adaptation of light wood construction, which has proved ideal in the historic centre; the use of lightweight solutions for soundproofing foundries; the use of traditional windows for internal ventilation in dwellings; the introduction of modern installations, which encountered no major problems; stucco technology; and a new approach to traditional formulae, roof space ventilation, etc.

All of which leads us to the conclusion that the best type of housing is one that can be constantly repaired and maintained, adapting to needs emerging during the inhabitant's lives. Such modifications can be made at a reasonable cost, which citizens wishing to continue living in these buildings can afford.

Environmental management

In the 1980s, the city's functions as regional capital and the increasing tourist influx began to increase the pressure of vehicular traffic. At the beginning of the period under consideration, one essential concern was to protect the historic ensemble from this pressure. People were encouraged to walk rather than drive, with new measures to shorten distances inside the city. A further objective was to reduce the density of road traffic in the whole city centre with diversions on to new ringroads.

The second major challenge was to modernize public transport. The major infrastructures (the international airport, handling 1.2 million passengers per year, the railway station with 2 million passengers, and the bus station with 6 million) are now linked to the city centre by public transport.

A large volume of regional traffic enters the city every day, mostly generated by the commercial activities in the large hinterland. This stems from the recent tendency to build residential areas in adjacent municipalities with less rigorous urban planning regulations, encouraging the destruction of its traditional type of habitat, and from the mass weekend movements of students. The transport structure, which until 15 years ago had consisted mainly of roads radiating out from the centre, has been replaced with peripheral ringroads taking in the traffic generated by the major supra-local infrastructures.

This road network is being extended with a non-toll-paying motorway, which has three exits to the city centre and handles traffic between the north and south of Galicia. An urban dual carriageway parallel to the motorway is also being developed to link the industrial estate to the north with the service area to the south, avoiding the centre. As stated above, the major public and private amenities and installations needed for a modern capital were built above this carriageway, including an industrial estate, cattle market, conference centre, shopping centres, sports complexes and hospitals. These are grouped around the historic centre, each of them equipped with very extensive parking facilities.

In the immediate environs of the historic centre, which are devoid of archaeological interest, 2,300 parking spaces were created, strategically situated to respect the functional characteristics of the historic city. These include the Juan XXIII coach and car park which plays a very important role in accommodating tourist vehicles. Nevertheless, given that the more car parks are built in the city centre the more traffic will come in, a second ring of car parks is being built providing 2,400 spaces. Also, further away from the centre, near the main roads into the city, there are 5,600 further parking spaces linking up with a bus shuttle service to the city centre.

In connection with daily traffic from adjacent municipalities, as part of the restrictions on municipal action imposed by regional regulations, a metropolitan agreement is being negotiated for the development of a combined public transport plan that will improve access and thus restrict the use of private transport.

Inside the historic centre, the municipality has initiated policies for regulating motor and pedestrian traffic. It is not merely a case of preventing movement of traffic, but rather of establishing timetables and guidelines for use, regulating times for loading and unloading lorries with a limit on vehicle tonnage. This is accompanied by special regulations on resident access once the barriers stopping general traffic from 11 a.m. onwards are raised. This policy requires constant negotiation with the users, since the effectiveness of the barrier system is liable to collapse if too many users are granted access and the density of traffic rises to its former high levels. Moreover, street furniture has been reduced to the bare minimum and strategically placed in order either to facilitate or prevent motor traffic or pedestrian use, as appropriate.

This has so far prevented any serious levels of pollution. Air quality controls are carried out systematically 24 hours a day in the area surrounding the historic centre where the traffic is heaviest, and the results have been satisfactory.

Tourism and heritage management

Over the last few years of the twentieth century, the municipality has realized that tourism, which had until recently been considered as an excellent source of economic activity, is taking on a mass dimension and therefore becoming a problem. The constant promotion of the Jubilee Years in 1993, 1999 and 2004 and the general cultural events, which will culminate in the year 2000, have led to a boom in numbers of pilgrims, tourists and visitors to Santiago de Compostela, the total for 1999 being estimated at 9 million. This seasonal influx completely disrupts the course of everyday life. The city is very much in vogue for many reasons. It has aroused interest throughout Europe and the rest of the world, especially among young people, by the theme of taking the 'Road to Santiago'. Above and beyond any religious reasons, this has to do with creating a new feeling of belonging to Europe and the modern desire for adventure. The city also has an indubitable quality heritage and its high degree of conservation, and also the ferment of activity that transforms it on a daily basis, has made it into one of the main landmarks of Europe. The private sector has responded well to the demand: small agreeable hotels are opening in the historic centre, and the major hotel chains are setting up in the areas set aside for the purpose.

However, tourism can be greedy, or even predatory, given that the heritage is neither renewable nor a consumer product, and indeed the first signs are emerging of damage to fundamental items, requiring urgent attention. We must move on from the mere 'ingestion' to 'digestion' of tourism, turning our backs definitively on the already discredited principle of 'the bigger the better'. In accordance with the guidelines of the plan, the city has been endowed with sufficient infrastructures and amenities to assimilate tourism, though this cannot be achieved as long as the avalanche continues to be concentrated in summer on the Cathedral and its environs. It is a case of pulling out all the stops in the city, the surrounding area and the whole region, by means of agreements between the authorities and other towns and cities. We must educate a professional sector that prefers to measure results in quantitative terms, and foster a dynamic view of tourism among the citizens. In a word, the organization must cover the whole geographical area throughout the year, applying sustainability criteria to tourist activities.

To this end, the municipality has developed a prestige tourist plan. Its aim is to resolve these problems, involving both the public and private sectors, and launching a public management enterprise with the participation of the private sector, to facilitate rapid, effective action. It is hoped that this will soon have an impact.

Sustainability

Concern about sustainability has been a constant feature since the original proposal to renovate the city. The urban development policy of the previous decades had resulted in a problematic situation in terms of mobility, environmental quality, population reproduction and diversification of activities in the historical centre. Therefore, one of the main objectives of the reorganization of urban growth and the programme to rehabilitate the built-up patrimony was restoring the balance among the different parts of the city and including once again, in the habitability guidelines, traditional elements related to the rational use of space and resources.

It is worth highlighting a number of aspects.

- The importance attributed to construction materials and solutions in projects involving built-up patrimony and, especially, homes in the historical centre, where improving the quality of life of the inhabitants is a priority objective.
- The suitability of having developed a new economic hub centred on a ringroad, avoiding excessive pressure on the historical quarter, which has enabled, among other things, the construction of a belt of 500,000 m^2 of green spaces around its perimeter.
- The problem of eliminating waste has been satisfactorily solved for a period of 10 years with the construction of a new processing plant, the only one in Galicia that meets European Union standards. By the time this plant reaches the limit of its treatment capacity, the second phase will have been put into practice and the region will have an integrated waste treatment system. The same is true of sewage; the capacity of the treatment plant has been enlarged in order to guarantee its effectiveness, even if the population continues to grow at its present rate.
- In social terms, diversified economic activity with a strong services sector has provided, despite the high unemployment rate that is endemic to Spain, an appreciable improvement in the level of employment.
- A professional and efficient range of social welfare services has managed to eradicate the pockets of shanty dwellings in the city and provides permanent programmes to deal with groups in danger of being isolated.

Conclusions

This chapter has considered general and specific planning, rehabilitation and development issues related to the city of Santiago de Compostela. The conservation programme for its architectural heritage has covered 12 years of integrated work that has involved an excellent technical and political team. However, it must be admitted that we encountered very many difficulties. It was a very complex process to secure political agreement on setting up the City Consortium, as was the attempt to introduce citizens to the rehabilitation mindset, accustomed as they were to an individualistic vision of trying to get whatever they wanted, usually at the cost of destroying the heritage. Furthermore, it was extremely arduous to introduce the pedestrianization policies and to block real estate operations aimed at exploiting the built heritage at the cost of destroying the social fabric.

However, once these difficulties were overcome, the city and the technicians also gained real satisfaction from their work. This is reflected in the number of major international prizes that have been awarded to the city. These include the Europe Prize of the Council of Europe, the Toledo Royal Foundation Prize, the Florence Prize, the Manuel de la Dehesa Prize, the Gubbio Prize and, in 1998, the Urban Planning Prize (a European Union award presented every four years to the city that has responded most successfully to its future challenges).

References

Council of Europe (1975): *Congress on the European Architectural Heritage, 21–25 October 1975.*

Kleihues, J.P. (1995): 'Urbanismo es recuerdo. Reflexiones en torno a la reconstrucción crítica', in Martí, C. (ed.) *Santiago de Compostela: La ciudad histórica como presente*, Consorcio de la Ciudad de Santiago de Compostela, pp.146–49.

Other sources of information

Costa, P. and Morenas, J. (1989): *Santiago de Compostela 1850–1950,* Colexio Oficial de Arquitectos de Galicia

Dalda, J.L. and Bardají, E. (1990) 'Santiago de Compostela: el Plan Especial', *La Ciudad como proyecto.* UIMP-MOPU, Madrid

García Iglesias, J.M. (ed.) (1993) *Santiago de Compostela*, Xuntanza, A Coruña.

ISIAC (1977): *Proyecto y ciudad histórica*, Colexio Oficial de Arquitectos de Galicia, Santiago de Compostela

Lópes Alsina, F. (1988): *La ciudad de Santiago de Compostela en la alta edad media,* Santiago de Compostela

Martí, C. (ed.) (1995): *Santiago de Compostela: la ciudad historicá como presente,* Consorcio de la Ciudad de Santiago de Compostela

OIKOS (1975): 'Notas sobre el desarrallo urbano de Santiago de Compostela en la década de los 60', *Ciudad y territorio,* 1/2: 95–106

David Lovie

Grainger Town, Newcastle upon Tyne, United Kingdom

Introduction

A conservation policy was first introduced to the City of Newcastle upon Tyne in 1961 and 'preservation areas' were defined in 1963, four years before the national advent of *conservation areas*. However, it was not until 1975 that conservation in the city received local recognition and support. The significance of 1975 is that it was European Architectural Heritage Year and the city's council and many of its inhabitants embraced what for them was to become a year long celebration of local architectural heritage through exhibitions, publications and other activities. These events coincided with a growing public feeling that redevelopment in the Newcastle of the 1960s and early 1970s had destroyed too much of historic and architectural value. This enthusiasm for an initiative of European standing was new for Newcastle, but it is now more than matched by the present thirst for European Regional Capital status and for European City of Culture designation (with Gateshead) in 2008.

In 1975 a landmark planning guide *Conserving Historic Newcastle* was issued. The city council was determined to develop conservation schemes, a *Town Scheme* of grant aid and positive conservation control of development as 'a firm foundation upon which work in future years can be based'. This renewed public concern and support for Newcastle's heritage was to be matched by the council's determination to develop and creatively exercise the necessary means and measures. Throughout the 1980s more sophisticated and better-targeted policies and grant aid schemes had a measurable impact on the level of vacancy and decay in the historic areas of the City. A 1987 conservation report (City of Newcastle upon Tyne, 1987) paved the way for a comprehensive city-wide Conservation Strategy to be developed in parallel with the new Newcastle Unitary Development Plan (UDP) that the council was required to prepare. Although council resources were becoming more and more limited – an annual conservation budget of £60,000 did not go very far with 1,600 protected buildings and seven large conservation areas in the city – a public consultation on the need for and contents of a strategy was carried out, and in 1991 a report identified ten conservation topics, four of which required urgent attention. Only one of these four topics *The Grainger Town Study* (1992) demanded that a territory (rather than a subject area) should receive early action. This was the first occasion when the *Grainger Town* name and improvement concept appeared on the public scene.

The policy and planning framework

Recording the physical condition of historic Grainger Town

Grainger Town covers approximately 36 hectares (90 acres) and contains over 500 buildings. This area includes all of the centre and northern quarters of Newcastle's walled medieval town, a town which celebrated its 900 year anniversary in 1980. Although the city has changed and spread well beyond its earlier area, Grainger Town still largely retains its medieval layout of markets, burgage plots, ancient radial roads and monastic precinct constrained by the remains of the military walls and dry moat. Some of this pattern was overlaid by the extensive and excellent early nineteenth-century development of builder Richard Grainger, which now gives the area its name and unique classical qualities (Fig. 13.1).

Although all Richard Grainger's grand terraces survive in some form, the full legacy of Grainger Town also includes many fine structures from both before and after Grainger, which together gives this quarter of Newcastle a

Fig. 13.1 Monument to Earl Grey (looking south) – this birds-eye view of Richard Grainger's formal Baroque plan of central Newcastle shows Grey Street on the left curving up from the River Tyne to meet Grainger Street at the huge monument to Earl Grey. (Photograph: City Repro, Newcastle.)

Fig. 13.2 Plan of Grainger Town – Grainger Town lies at the heart of the city of Newcastle upon Tyne and contains many high-quality protected historic buildings. In 1996 nearly half of them were considered to be at risk from neglect and decay. (Courtesy of Ordnance Survey.)

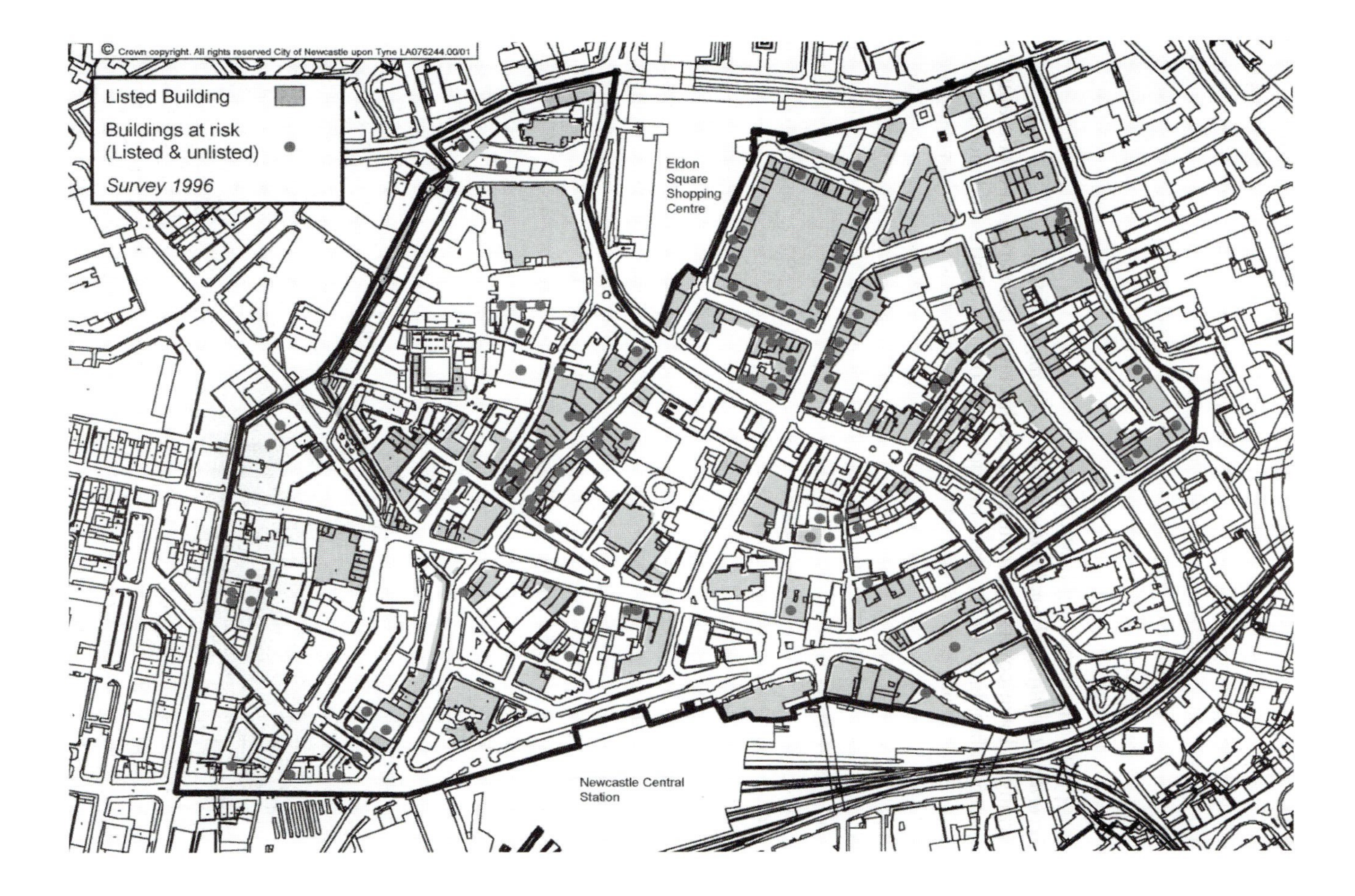

national, if not international value. In fact 40 per cent of the buildings contained in Grainger Town are included in the English National *List of Buildings of Special Architectural and Historic Interest* (i.e. they are afforded statutory protection). Moreover, 20 per cent of these have Grade I status which confirms this assessment of quality (only 4 per cent of the country's total stock of listed buildings have this level of status). But by 1992 the vacancy and physical condition of both buildings and the spaces between them gave much cause for concern (Fig. 13.2).

The *Grainger Town Study* (1992), commissioned from a consortium of consultants in 1991 by a partnership of the City Council, English Heritage (the Government's National Conservation Body) and Department of the Environment (a Government Ministry acting through the regions), identified that about one-third of the floor space in the study area was vacant and unlikely to be reoccupied without substantial investment. As the study said, 'this unoccupied 1.7 million square feet (or 158,000 square metres) is sufficient to house nearly 7,000 people or provide working space for some 11,500'. This vacancy, some of it long-term, had led to widespread neglect, serious building decay and a consequent run-down appearance.

To support these findings, a detailed survey of the condition and vacancy of all Grainger Town buildings was carried out according to English Heritage's national criteria of *buildings at risk* whereby buildings are categorized according to the degree of vacancy and extent of decay. From this comprehensive survey, 47 per cent of the listed buildings had severe problems of decay and vacancy while a further 29 per cent were in a marginal condition making them vulnerable to further neglect and decay. The equivalent national figures are 7 per cent and 14 per cent respectively (Lovie, 1995). Although building problems were scattered throughout Grainger Town, there were particularly bad concentrations around Grainger's market and Grainger's lower quality streets. It is clear that the players of the 'property market' in 1992 for crude or ready retail and commercial space did not favour Grainger's Georgian terraces, the original upper floor residential population having moved out to the suburbs long ago.

In carrying out and later updating the buildings at risk survey, three particular problems were encountered. First, surprisingly, it can be difficult when dealing with terraces and previous multi-occupation and mixed buildings from different periods, to establish the extent of each building to be surveyed and recorded as a unit. Second, it is seldom possible to gain access to the interior of every property for survey, meaning that many judgements can only be made from external indicators. Finally, if different surveyors are used, there may be inconsistencies in judgement of condition, in spite of the use of common indicators. A pragmatic approach to all three problems usually provides a reasonably reliable result. In the case of Grainger Town, there was no mistaking the seriousness of the neglect and decay. Furthermore, it was acknowledged that the area lacked open green spaces and that the existing spaces between buildings were poor in contrast to the very high quality of the buildings. Very little was worthy of an aspiring European Regional Capital of the importance of Newcastle.

The preservation, enhancement and design issues

In addition to the issue of poor physical condition of buildings and spaces, Grainger Town was also suffering other less tangible problems, which were probably the causes of the physical problems too.

Following the loss of the residential population in the late Victorian period, upper floors were taken over for offices, workshops and storage. Many of these uses, in their turn had also diminished or moved on, and with the movement of the centre of retailing activity away from the Grainger Street area, ground floors as well as upper floors were becoming vacant, or else occupied on unsatisfactory short-terms lets.

A major deterrent against natural economic regeneration was that the 'hope value' given to the properties by owners was well above the realistic value for rental or for sale. This sets up a vicious circle which can only be broken by encouragement of a variety of uses supported by public investment to aid the actual (not the perceived) shortfall in value.

Furthermore, expectations for the availability of operational and support parking were too high. Too many assumptions regarding the necessity for parking spaces in all changes of use on the part of local property professionals, has proved to be a particular problem. Because ideal parking circumstance could not be offered in most cases, local attitudes to the viability of re-use were prejudiced. Experience and examples of re-use, especially residential, have since demonstrated that there is a wide diversity in the market regarding the necessity of car parking. Having said this, there is still a reasonable need to provide some parking spaces.

Finally, the pedestrian environment has been much eroded by the growth of traffic, which has resulted in noisy, polluted, intimidating streets and spaces. Some routes have become so heavily trafficked that they are effectively a barrier to pedestrian movement particularly between Grainger Town and the southern city centre.

Essentially, sustainable regeneration will only take place if physical regeneration is linked to an effective strategy for economic development. It is essential to establish the best balance between Grainger Town's economy and people, and its unique physical sense of place. Should place be overemphasized then the necessary development that can sustain the existing place will not happen. The implications of this is that preservation should be limited to the precious medieval structures in Grainger Town, while the remaining valuable buildings should be conserved to retain essential character and features. This is not a sacrifice of principles, but a careful and creative recycling of heritage to enable it to earn its living and so to survive into a viable future without losing its quality and integrity.

Physical enhancement of the streets and spaces with improved surfaces is a necessity in Grainger Town, but dynamic influences such as traffic should also be tackled, which is more difficult as solutions may lie partly outside Grainger Town. Also, such movement should be reduced in a way that does not damage existing and developing business. This could take time to enable attitudes and expectations to adjust to the greater benefits of a pedestrian-dominant environment.

Fig. 13.3 Grey Street (looking south) – this beautiful 1830s street on a rising curve has been long considered one of the finest architectural streets in the United Kingdom. (Photograph: City Repro, Newcastle.)

Fig. 13.4 The Union Club overshadowed by Westgate House – the 1960s brought these two unlikely neighbours together. The 1877 Union Club has been restored and is now a well-loved public house. Westgate House is now a target for demolition by the Project. (Photograph: Grainger Town Project.)

The issues for design are the usual familiar ones of achieving compatible scales, materials, uses, etc., but in the Richard Grainger streets with their pleasing uniformity, conformity and control must be strict so as to maintain the integrity and quality of the area (Fig 13.3). In the few places where new build is possible, new development should appear as a natural evolution of the townscape compatible with the usual incremental nature of change in historic towns. This has not always been achieved in the past in Grainger Town, so opportunities should be taken (or made) to reverse or remove these previous unhappy interventions and to replace them if necessary or appropriate, with more suitable developments. This is going to be possible with only one such 1960s building, the monstrous Westgate House office block, because of the time limits of the current project. This removal will not only dispose of an eyesore but will also help to greatly enhance the settings of nearby historic stone buildings (Fig. 13.4).

The integration of urban planning and conservation

Because of the joint origins of planning and preservation in the United Kingdom, integration of urban planning and the historic built environment has long been a feature of statutory activity and official guidance since 1947. This is particularly true at a national level, with an unbroken link to the current *Planning Policy Guidance: Planning and the Historic Environment* (PPG15, 1994) via legislation: Historic Buildings and Ancient Monuments Act, 1953; Civic Amenities Act, 1967; Town and Country Planning Act, 1971; National Heritage Act, 1983; Circular 8/87, Historic Buildings and Conservation Areas – Policy and Procedures; and Planning (Listed Buildings and Conservation Areas) Act 1990.

At a city-wide level, government guidance on the preparation of Unitary Development Plans (UDP) has always stressed the integration of planning and conservation. This is true of the Newcastle UDP, adopted in January 1998, in which building and area conservation is both a strategic planning objective as well as an integral part of local planning achievement. The Newcastle UDP also contains a comprehensive range of policies on Listed Buildings, Conservation Areas and Archaeology to be applied throughout the city, including Grainger Town. Not only does Grainger Town contain about 250 *listed buildings*, but most of it is within the Newcastle *Central Conservation Area* and all of it within the *City Centre Area of Archaeological Interest* as designated in the 1998 UDP. Consequently the UDP's conservation planning policies apply directly to Grainger Town and they cover:

- preservation of the built heritage;
- alteration or extension of listed buildings;
- demolition of listed buildings;
- façadism;
- development in conservation areas;
- demolition of unlisted buildings in conservation areas;
- development that would harm areas or sites of archaeological interest;
- archaeological assessments;
- preservation *in situ*;
- accidental discovery.

At the level of a single *conservation area* (similar to Grainger Town – which forms part of the Newcastle Central Conservation Area), the type of specific action plan for conservation that the government recommended to local councils in 1987 are *Protection and Enhancement Plans*. These enable the authority to take a comprehensive approach to the proper cultivation of a conservation area into the future.

This comprehensive protection and enhancement approach was taken to the future conservation and planning of Grainger Town by the report of the *Grainger Town Study* (1992). The objectives of the Study were:

- to examine the problems and causes of under-use, fabric decay and environmental erosion by traffic;
- to identify regeneration options that balance traffic needs and economic revival based on new roles, compatible with conservation of its historic fabric;
- to identify a preferred option that is practical and affordable for the council in partnership with the private sector.

By putting conservation in context with economics and traffic, and by championing public/private partnership working, it paved the way for the more detailed reports that followed.

In 1994 *Grainger Town Conservation Area Partnership* (CAP) was established on the basis of an agreed *action plan* (Lovie, 1995). It was one of a small number of partnership funding schemes operated jointly by the local councils and English Heritage. However, this funding mechanism has since been superseded. Nevertheless, the early actions and further reports served to guide and promote the setting up of a formal Grainger Town Company and Partnership that would employ a full-time *Project Team* to pursue the regeneration of Grainger Town over 6 years, beginning in 1997. These reports are as follows:-

- *Grainger Town Regeneration Strategy* (EDAW, 1996);
- *Grainger Town – Working for a New Living* (Newcastle City Council and English Partnerships, 1997);
- *Grainger Town – Vol. III: Public Realm Strategy* (Gillespies, 1998).

The use of design/development site briefs and design guides

The early preparation of site development briefs was essential to the proper establishment and continuing support of the Grainger Town Project. Such comprehensive briefing initially gave credibility to the Project and greatly aided the search for developers and facilitated early programming of resources and staff effort that had a high measure of reliability.

The briefs provided by the three detailed reports mentioned above covered all the key sites in Grainger Town in varying degrees of depth. In total, the reports provided:

- development potential studies for 5 sites;
- development frameworks for 6 areas;
- building development briefs for 4 sites;
- environmental improvement briefs for 8 areas of the public realm.

Fig. 13.5 New shopfronts around Grainger Market – The 'shopfronts' grant policy is aimed at securing an imaginative blend of old and new – to provide a suitable role for the past in the new Grainger Town. (Photograph: Grainger Town Project.)

In addition, they supplied overall development strategies, land-use strategies, urban neighbourhood divisions and an urban design framework.

Some formal and informal design guides had been inherited from the City Planning Department and usually covered a larger area than just Grainger Town. An informal sign guide was developed for parts of *Grainger Town*. A shopfront design guide (Fig. 13.5) was produced in 1996 for use in Clayton and Grainger Streets (which with Grey Street form the main focus of Richard Grainger's planned redevelopment of the city in the early nineteenth century) and will shortly be updated. A separate guide was developed to upgrade the quality and restore the 220 stalls in the Grade I listed Grainger Market and Arcade.

The Grainger Town Project has also embarked on the preparation of what can only be called *Attitude Guides* aimed at fostering appropriate *conservation attitudes* in property owners and investors and their professional advisors. The first, called *Investing in Quality*, recently produced for the Partnership by URBED, a national regeneration and communication consultancy, promotes the real value of historic and architectural quality and, it will be launched in 2000. This will be followed up by guidance for prospective owners of listed buildings and, specifically, on how to properly maintain them. The Project is convinced that influencing attitudes will provide more enlightened owners and professionals who will make better development partners with the Project and the city council in the future.

The *Gillespie Report* (1998) also proposed two key projects which could greatly enhance the streets and spaces of Grainger Town. The consultants suggested that a 'lighting strategy' be developed to create feelings of both drama and repose as is appropriate, and an overall public art strategy to reflect and embellish the pleasures of Grainger's Streets.

These guides and briefs are supported by demonstration projects carried out on the ground that show how to achieve the best results. Such projects include the conversion of the Central Post Office (creative mixed uses), Clayton Street (residential conversion of upper floors), Grainger Street (creative shopfront design), and Phase 1 of the Public Realm Strategy (high quality, floorscape).

The integration of new developments

This is probably the most difficult part of the process of cultivating historic environments. The dilemma is that conservation areas need the injection of new development that represents new investment, but too often in the past the results have been aesthetically unsuccessful and controversial, resulting in a loss of face for the responsible authority. Where we do see what are believed to be successful integrations, they have either been developments behind retained façades or pastiches of history or else very expensive modern buildings.

The option of façading in Grainger Town has little public support and is anyway, heavily discouraged by Newcastle's UDP (Policy C2.2), particularly of Richard Grainger's buildings, preference being for retaining the whole original range of rooms. Pastiche has some support as in the recent replacement of a collapsed historic frontage and the approval given several years ago to a pastiche replacement of 2–12 Grey Street and not yet carried out. This latter option will be used in a limited way to help to restore the integrity and unity of Grainger's streets. Elsewhere in Grainger Town the story is more mixed.

The 1960s and 1970s saw a number of inappropriate intrusions into the historic environment such as the oversized Westgate House. Although it is unlikely that such developments would receive planning permission today – in fact the Grainger Town Project is on track to demolish this eyesore to heal this long-standing wound in the townscape integrity – recent examples of new build in these areas have not been entirely satisfactory.

In spite of the general, well meaning guidance in the UDP and the design guidelines and site development briefs from consultants to the Grainger Town Project, appropriate and even good design for new build cannot be guaranteed. It still requires several difficult and often subjective intangibles to be put in place. These include: an enlightened developer, a skilled and imaginative architect/designer, the capacity of professional and lay participants in the development control process to recognize quality and appropriateness, an unwillingness to compromise and sufficient funding. These factors will assist in producing schemes that fit into the historic environment through creative subservience, harmony or contrast. None of these intangibles can be guaranteed but the Project is deliberately setting out to help to create them by:

- being a formal partner in the city council's development control process;
- having an Urban Design Panel of experienced local professionals to advise the Grainger Town Board on the design aspects of planning applications;

- having Resident and Business Forums whose views are invited (and valued) on planning applications; a commendable emphasis on 'conversational democracy';
- employing full-time heritage and development officers experienced in design matters and in the intricacies of the statutory development control process;
- publishing design guidelines to establish standards for conversion and new build;
- publishing the booklet *Investing in Quality* to influence owners, developers and professionals;
- providing adequate funding to our own schemes and grant aid to private schemes to help to increase the chances of properly funded, high quality design solutions.

All this has already begun to pay off through mixed restoration/new build schemes and the 'Public Realm' floorscape improvements around the landmark Grey's Monument. It is also too early yet to pass judgement on the ability of the decision-making process to produce quality with respect to the current major proposals for new buildings in Grainger Town. Moreover, the 6-year life of the Project is a little too short to make it clear that lessons in integrating new buildings have been learned as a result of the Project; only a longer time will tell – once the projects are completed and functioning.

Management and regeneration action

Regeneration via conservation-led strategies or other approaches

The success of any urban regeneration project, whether conservation-led or not, is increased in proportion to the number of urban development factors that are brought under control. In simple terms, a project that sets out to only regenerate building fabric will be less successful than one that also sets out to support building uses, such as residential or commercial. The Grainger Town Project, although conservation inspired, set out from the start to address a variety of factors in the regeneration of the area. The regeneration of any complex historic urban system needs a multi-layered approach that works with and not against the grain of historic and architectural character.

The *Working for a New Living* (1997) report presented the factors that the Project was to address as action orientated 'Regeneration Themes'. There were seven of them described in the report as follows:

1. *Quality of environment* – to improve the quality of the public realm and traffic management and create an attractive environment which promotes confidence in the area.
2. *Business development and enterprise* – to develop existing businesses and promote the formation of new businesses.
3. *Housing* – to increase the residential population by creating a range of affordable housing for rent and sale.

4 *Non-housing property development* – to secure the re-use of historic buildings and the redevelopment of key sites for office, retails, arts and leisure use.
5 *Access to opportunity* – to improve training and employment in Grainger Town for residents in the adjoining inner city wards.
6 *Arts, culture and tourism* – to promote Grainger Town as a centre for the arts, culture and tourism.
7 *Management, marketing and promotion* – to improve the overall management and marketing of the area.

This list of factors has been criticized because 'conservation' as an activity and goal *per se* is not mentioned. But because the area is of such high heritage value with a high level of protection built in, and because the project was originally conceived because of the need to actively conserve Grainger Town, conservation is a thread that runs irresistibly through every one of these 'Regeneration Themes'.

The 'Vision for Grainger Town', established in the same publication, quietly but positively testifies to the value of conservation in the regeneration process. The vision is as follows:-

> Grainger Town will become a dynamic and competitive location in the heart of Newcastle. This historic area will develop its role in the regional economy, within a high quality environment appropriate to a major European Regional Capital. Its reputation for excellence will be focused on leisure, culture, the arts and entrepreneurial activity. Grainger Town will become a distinctive place, a safe and attractive location in which to work, live and visit.

Conservation is clearly necessary in 'this historic area' to maintain its 'high quality environment' as a 'distinctive place'. The value of presenting conservation in this gentle way, is that it is seen in comfortable harmony with developing 'a dynamic and competitive location' and 'its role in the regional economy' through 'entrepreneurial activity'. In short, this avoids the perennial, time-wasting and crudely divisive arguments about *conservation* versus *development*. The Grainger Town Project contends that in historic environments, *conservation* needs *development* and *development* needs *conservation*. This, of course, does not remove all conflict, but it goes a long way to make that conflict creative for both *conservation* and *development*. In fact, our publication *Investing in Quality* (URBED, 2000) promotes the positive value of historic buildings and their role in powering regeneration.

The role of national and local authorities and agencies
The origin of the Grainger Town Project lies in 1991 when Newcastle City Council accepted a proposal from English Heritage to share in commissioning a study of the northern part of the historic centre of Newcastle, subsequently called Grainger Town. So from the beginning, the basis of

Table 13.1 Members of the Grainger Town conservation partnership, 1999

Public partners	*Status*
Newcastle City Council	Local authority
English Heritage	National agency
Northern Arts	Regional agency
Northumbria Tourist Board	Regional agency
Tyneside Training and Enterprise Council	Local agency
English Partnerships (ONE North East)	Regional agency
Private partners	*Description*
The Newcastle and Gateshead Partnership	Private and public body
North East Chamber of Commerce	Regional trade body
North East Civic Trust	Regional conservation body
Entrust	Local enabling organization
Project North East	Local enabling organization
Bowey Group	Local building company
Boots Ltd	Local/national chemists
Chestertons	Local/national property agents
Dickenson Dees	Local law company

the project was to be a partnership initially between a local authority and a national government agency, each with the responsibility for conservation at their respective levels. The partnership for the study was then joined by the Department of the Environment, a national government department, working on inner-city regeneration in the north-east region of England.

For the management of the Project and to spread its effectiveness, the initial partnership of three was greatly extended further into the public sector, but also, for the first time, into the private sector. Table 13.1 shows a breakdown of the current partners involved in managing and funding the Project through a Board of twenty and an Executive Board of ten:

Although each one of these partners have a separate and special role in their own fields of operation, they all share in the common role that is played by the Grainger Town Board, i.e. responsibility for matters of Project policy, priorities, strategy and appraisal, and approval of all schemes, administration, monitoring and evaluation. In addition, Board Member's skills and experience developed in their own special areas of operation, are made available to the Project Director and his staff for the benefit of day-to-day Project work, as well as on the consultative forums and panels.

The establishment of management agencies for the regeneration of historic centres

Following a review of regeneration agency arrangements in other United Kingdom major centres of regeneration such as Leeds, Liverpool and

Glasgow, consultants recommended that the legal entity best suited to the situation and purposes of the Grainger Town Project would be an independent company limited by guarantee. In particular, this arrangement was considered to be more flexible allowing for more accountability and effective tailoring of schemes, would enable the powers of the local authority to be more readily available and would not require a lengthy legal process to establish it. Essential to the effectiveness of this Partnership Company would be a structure that include a Board, an executive team and a consultative forum. This pattern has essentially been followed and each element is examined below.

The executive team is now called the Project Team, and is made up of fourteen officers. The tasks involved are varied and complex, so the Team is multi-skilled in order to address the requirements of the 'Regeneration Themes'. The Team is divided under two Assistant Directors. One side deals with 'Physical Development' and includes several Development Officers and the Heritage Officer, whilst the other side called 'Projects and Monitoring', has three specialist Development Officers – one each for Employment, Business and Culture – and a monitoring Accountant and Information Officer. Of course, in everyday working, officers work in a complex multi-disciplinary way in which their varied specialist skills and experience are woven together to produce the desired results.

The Team is led by a Project Director who is appointed by the Partnership Board and is line-managed by Newcastle City Councils' Director of Enterprise, Environment and Culture, an arrangement that has proved to be very satisfactory. The Team is completed by an Office Manager and administrative support staff.

To assist the Partnership Board and Director, the following consultation and advisory groups have been set up:

- the Grainger Town Residents Forum;
- the Grainger Town Business Forum;
- an urban design panel;
- an arts and culture panel;
- a housing steering group;
- a training and employment steering group;
- a business grants panel;
- a cultural grants panel.

They all report to the Board and the forums receive Board papers to comment on. Administration is kept to a minimum and attendance at the Residents Forum and other meetings and events involves staff in out-of-hours working.

The nature and relationship of funding agencies for conservation and regeneration and their impact on the cultural built heritage

Due to the fact that the Grainger Town Project covers a wide range of activities, it has pushed regeneration action forward on a very wide front, wider than many previous regeneration projects in the United Kingdom.

Table 13.2 Grainger Town Project: core funding agencies

			Purposes	
Agency	*Status*	*6-year contribution*	*Conservation*	*Regeneration*
Newcastle City Council (NCC)	Local	£2m	Grants to repair buildings at risk and improve shopfronts	
English Heritage (EH)	National	£1.75m	Heritage Economic Regeneration Grants share the above work with Newcastle City Council.	
Tyneside Training and Enterprise Council (TEC)	Local	£0.25 m		Employment Development and Training.
English Partnerships (EP)	National	£25m	.	Grants to convert or refurbish housing or offices. special develop-Schemes
Single Regeneration Budget (SRB)	National (Government)	£11m		Very wide range of improvement schemes covering employment, training, business, public realm, housing, shop-fronts, ethnic minorities and crime and safety. Employment of three specialist Development Officers.
	Total	£40m		

Consequently its funding base is also very wide, drawing from local, national and European sources.

The Project's core funding of about £40m for the whole of its 6-year time-scale comes from five separate agencies, two of which supply specifically conservation funds whilst the others provide general regeneration funds, some of which are incidentally used for conservation purposes too but in a more generalized improvement context. In particular, the regeneration funds serve 'conservation' by encouraging and developing new uses for historic buildings. Both local and national funds are used for both purposes as the comprehensive funding Table 13.2 shows.

The application of the funds supplied by the core funding agencies illustrate the extraordinary variety of activities which, in creative combination, are already bringing about the regeneration of Grainger Town.

Newcastle City Council (NCC)
As the local authority, the City Council has direct responsibility for conservation in its area. The £2m given to Grainger Town has been offered in joint grants of between 60 per cent and 80 per cent with English Heritage to private owners or occupiers of vacant and/or very decayed historic buildings, to repair them as well as to return disfigured shopfronts to a more elegant, traditional appearance. Since the grant scheme began in 1994, 114 buildings have been rescued from decay and 59 shopfronts improved.

English Heritage (EH)
As the government's national agency for conservation and heritage, English Heritage carries out a wide range of tasks including managing historic properties for tourism, directing and funding archaeological investigation, advising on the development of private and public historic buildings and monuments and, as in the case of Grainger Town, supplying funding for the repair of historic buildings and shopfronts. This joint scheme with Newcastle City Council has recently been described as 'one of the largest regeneration schemes in which English Heritage has been involved, and it has achieved some of the most spectacular results' (English Heritage/Town Centres Ltd/London School of Economics, 1999). This joint grant aid scheme was the first funding programme in Grainger Town and has not only rescued properties, it has also proved to be the catalyst for very much more public and private investment from a wide range of agencies. To date every £10,000 spent by public agencies on conservation in Grainger Town, has levered at least £50,000 of matching private investment, and in several cases very much more.

Tyneside Training and Enterprise Council (TEC)
This is one of a network of Training and Enterprise Councils set up in disadvantaged areas to develop vocational training initiatives and encourage local business enterprise. Clearly developing such opportunities in the centre of the city will help to bring new life to Grainger Town and in particular, will assist in generating and developing new uses to occupy vacant upper floors of historic buildings, thus serving the ends of conservation too. The TEC has a well-established network of contacts and funding available, both of which are particularly helpful in serving the purposes of the project.

English Partnerships (EP)
This is a national development agency operated through regional offices. The local office has now been absorbed into the Regional Development Agency and has replaced its Partnerships name with *ONE North East*. This

agency, which has a seat on the Board, has long regarded Grainger Town as one of its flagship schemes in the north-east of England. Its willingness to provide over half of the Project's public funds indicates its faith in the need and potential of the Project. Its £25m contribution is available as shortfall grant aid to assist private owners refurbish their upper floors for new residential or old office uses. In all, ten key development schemes have been targeted in the area as possible recipients for development aid and to increase the viability of the scheme or to raise the standard of its design. The scheme to demolish the 1960s Westgate House and replace it with more suitable development in scale with its historic surroundings, is one of these key schemes.

Single Regeneration Budget Challenge Fund (SRB)
This is a national government fund set aside specifically for urban regeneration to which local authorities make competitive bids.

Newcastle was successful in its bids for funding in 1995, the first year of the fund (SRB I) and in 1997 (SRB III). Of all the funding sources successfully engaged by the Project, this one is able to provide support for the widest range of regeneration activities. Although its purpose is to support economic development it is able to provide both capital funds to support infrastructure improvements and revenue funds to support training, employment initiatives and delivery staff. All these activities are aimed at improving the environment and bringing new economic life into the regeneration area, two aims which are capable of both directly and indirectly assisting in the conservation and reinforcement of the historic and architectural character of Grainger Town. The best way to demonstrate the wide range of basic and imaginative activities possible under this funding source, is to list the Grainger Town operations currently funded under SRB III. They can be divided into capital and revenue funding, and the main aspects of each are listed below.

CAPITAL

1. Grant aid to property owners to:
   - improve shopfronts;
   - convert vacant upper floors to student, private, or housing association housing (a 'Living Over the Shop' scheme).
2. Good design and improvement of public realm: floorscape, lighting and street furniture.
3. Establish and help implement a programme of commercial non-housing developments (with English Partnerships).
4. Provision of security cameras to combat crime.

REVENUE

1. Support for the running costs of Grainger Town Project Delivery Team, including the management function and the costs of three specialist officers for business, employment and culture.

Fig. 13.6 Conversion of the 1874 Post Office – a grant of £1.66 million made this imaginative £5.6 million conversion from a post office to a multi-use building possible in 1999. (Photograph: City Repro, Newcastle.)

2 Support for various innovative training initiatives, targeted at long-term unemployed to improve employment opportunities for them in Grainger Town (run jointly with Tyneside TEC). Funds also to provide training and support for an imaginative scheme called 'Newcastle Knights' which supports young people as local uniformed guides and urban rangers.
3 Support for business development of both existing and new businesses. 'Project North East' – a local training and enterprise agency – is involved with developing business opportunities for young people. A training programme called 'Welcome Host' to improve presentational skills in retail and other services in the area.
4 Support for existing and new cultural businesses and a programme of cultural events including street performances to attract people into Grainger Town.
5 Support for an initiative to encourage diversification within the local economy and create employment opportunities to benefit ethnic minorities.

The £40 million funding derived from the core funding agencies is expected to attract at least £80 million of investment from the private sector. (Fig. 13.6).

Initially, with such a large range of different funds there were problems of overlap, contradiction of grant aid 'conditions', lengthy approval times and confused customers. As time has gone on, and officers and agencies have become more used to joint working, the problems have sorted themselves out with a little negotiated give and take. Unfortunately, some agencies have attempted to clarify processes by increasing the necessary paperwork which has made customers more and more dependent on officers to see them through. This is not entirely desirable, but, with a little streamlining here and there and some patience in the face of occasional delay, the system can be made to produce the desired result.

What officers fear most in the face of such paperwork is that potential participants will be discouraged from joining in and rather than 'going-it-alone' without financial help, will prefer to do nothing and buildings will further decay and opportunities within the short life of the Project will be lost. If the 'carrot' of funding should fail, the Project will find itself forced to draw the city council into using its 'stick' in the form of statutory *Repairs Notices* or *Compulsory Purchase Procedures*. Although the Project would not hesitate to request the use of the stick should it be necessary, because its use can be expensive and time consuming, there is a preference for resolving matters at the carrot stage, which gives the Project the chance to be pro-active, as well as reactive.

There are also other supporting funds that have been attracted from public sources:

- Newcastle City Council
  - funding of five full-time officers in the Grainger Town delivery team.
  - substantial contribution to 'public realm' floorscape improvements.
- Heritage Lottery Fund
  - funding of about £100,000 to support the repair of Grey's Monument, a landmark and Grainger Town icon.
- European (totalling in excess of £1.8 m)

  European Regional Development Fund includes:
  - funds to support business development, job creation and cultural developments.
  - a grant of up to £600,000 to restore eighteenth-century Alderman Fenwick's House, Pilgrim Street.

  European Social Fund:
  - includes funds to support training and Youth Enterprise developments.

The relationship between commercial and social needs within the historical environment

Historically, most European urban centres have had a tradition of multi-use buildings, where domestic, social and commercial uses co-existed side by side. Whilst this tradition has continued to thrive in many countries, it has not been the case in the United Kingdom. As a result of years of land-use planning and the residential desertion of the town centre, our towns and cities have increasingly become simply collections of areas of mono-use, such as transport corridors, housing estates, industrial estates and shopping precincts. This has resulted in antipathy towards the historic town centre, which, at worst, becomes a tourist ghetto.

Experience of the enduring urban qualities of such European cities as Florence, Paris and Amsterdam, which have continued to feature multi-use activity and have mutated and evolved through the centuries, has had an impact on British visitors in recent times. There is now a growing belief that the British historic centres can be enjoyed for their own sake, and can also make suitable residential environments. However, the potential to create living units are unlikely to cater for families as the infrastructure to support them in the centres such as schools, have mostly disappeared.

The movement back towards the town centre is likely to be made up of mainly the young or old in small one and two person units, except for large units (or even mass accommodation) for students. Such an influx is already making a difference to Grainger Town. No longer does the centre close down after offices close in the evenings or at weekends when the shoppers have gone. There is now a growing city centre community with a vested interest in the appearance, cleanliness and all-hours safety of the centre. This is likely to result in a continuing improvement of the quality of centre public spaces and of the growth of a varied city centre culture, apart from the benefits of efficient use of buildings through multi-use occupation.

But this move towards the desirable residential/multi-use centre has not been achieved without conflict. In Grainger Town some new residents have had difficulties coming to terms with commercial and leisure uses nearby that have different time scales and noise practices than are found in the familiar residential suburbs. It is with this problem in mind that the environmental health aspects of locations are considered, and, for example, student accommodation is more likely to be supported in the noisier locations than housing association flats. In Grainger Town's Bigg Market area which has become the centre for intensive and lively leisure activities, the Project prefers to encourage student residential accommodation on the upper floors above and next to the many bars and pubs. It is all a question of striking the right balance of uses and users.

The over-riding commercial need, which is expounded by most of Newcastle's property agents, is operational car parking. This is a complex problem, which will be looked at later in the context of Environmental Management (see below).

The need for flexibility in the use and re-use of historic buildings and the relationship between value and vacancy/occupation

Except for the threats of demolition, the adaptation of historic buildings for new uses is the most difficult task faced in the conservation of our historic environment. For some urban professionals there are just two simple (and crude) solutions.

1 *Façadism*: reduce the historic building to its façade and build a new building behind it. This is based on the belief that 'façades' are what

give the buildings their character – and it is a good way of getting the best of the old to go with the best of the new.

2 *Maximum reversibility*: carry out all the required adaptations in such a way that they can be easily removed at a later date. This is clearly less damaging but can lead to too many adaptations that completely compromise the integrity of the building for the whole time the adaptations survive. Also, at a later date, a lack of information or progress of decay may make it difficult (or undesirable) for the work to be reversed, so the compromise becomes permanent or distorts the building's history into the future.

So, putting aside such simple solutions, the conservation authority must take more difficult routes towards a solution. There are a few guidelines from experience, which can help the process towards a logical conclusion, as follows:

- It is always wise to question the need for change. This may be difficult when a regeneration project only tolerates conservation when it does not get in the way of achieving serious and measurable targets of businesses created, jobs created and floorspace improved. But this questioning may not lead to refusing change, but rather to understanding its origins and it being accommodated elsewhere or in a less damaging way. Such dilemmas do not usually arise with precious ancient or highly graded buildings.
- Ideally, change to a property should best be thought of and designed as a natural evolution of the original building. This standpoint is more likely to result in changes that are in scale and character with the historic building and its environment.
- When faced with no choice – usually for circumstantial reasons – but to accept unsatisfactory alteration, it is important not to lose the will to win some improvement, as all schemes are capable of some successful modification which will be acceptable to all. Treat each case on its merits and hold onto principles and try to reduce the influence of circumstantial requirements.
- The key aspect is to find suitable uses for listed buildings, rather than be overly prescriptive, i.e. go with the grain rather than fighting against it.

Although it is important that a positive approach to change is adopted because for each new generation historic buildings have to continue to earn their keep, large-scale adaptation is not inevitable or adaptation necessarily bad. Considerable help in assessing the flexibility of buildings can come from the very nature of the building itself.

For example, in Grainger Town the many late Georgian-palace-fronted terraces built by Richard Grainger himself are fairly simple 'egg-crate' structures with a certain flexibility of character and integrity already built in. Also, because of their uniformity and repetition, strategies of trading-off changes to several already damaged houses for one or two fully restored

Fig. 13.7 Conversion of 1901 Galen House – formerly pharmaceutical offices and manufacturing, Galen House was converted into 62 flats to provide living accommodation in 1998. Both the conservation and commercial Grainger Town grants helped to secure this £2.5 million transformation. (Photograph: City Repro, Newcastle.)

properties, are possible. This is a legitimate exercise of principle in the face of aggressive reality.

Industrial buildings with large floor plates offer flexibility in use. For example, a large office/workshop complex that served the pharmaceutical industry from 1901 under the name of Galen House in Grainger Town's Low Friar Street, was successfully converted into sixty-two flats whilst maintaining the basic integrity of the building (Fig. 13.7). The Bee-Hive Hotel in Bigg Market went from hotel to unsuccessful offices before being converted back to residential use for students with grant aid from the Project. Having the largest 'Living Over the Shop' grant scheme in the United Kingdom, the Grainger Town Project has supported many satisfactory conversions of *listed buildings* whose integrity and future life has been enhanced by exploiting their basic flexibility.

The issue of 'hope value' (discussed above) has played a large part in the fall and rise of Grainger Town. In the economic booms of the 1960s and 1970s property dealers and owners put together mixed portfolios of properties from different towns and cities for direct sale or auction as the rush for central urban property for redevelopment attracted more and more investors. Unfortunately, the value of many of these Northern portfolios offering valuable sites in Leeds or Manchester, was padded out with single or double terraced properties in Grainger Town. Often the purchaser of these portfolios only visited the more valuable properties in the portfolio, leaving the Grainger Town properties to be purchased unseen or unsurveyed. Many of these properties looked in better condition in photographs than they were in reality, resulting in properties in dubious condition being overpriced in the portfolio and the owner expecting this high value from the market. Thus, there developed a considerable discrepancy between the expected (or 'hope') value and the actual value to be achieved in rental income or from sale. Many owners therefore hung on to their properties, spending little on repair and allowing upper floors to become vacant in the 'hope' that the market would change and they would eventually get their money back, possibly with interest.

This 'hope value' leads to a spiral downwards from a stable market as unimproved properties held against better times, blight the whole area, making such better times less and less likely. Although many cities suffered in this way, the very multiplicity of small ownerships of Grainger Town's many grand terraces by absentee landlords, meant that it was more vulnerable than most centres. Changing retail trends and locations in the 1970s also helped to intensify the blight, but the main underlying problem with many properties was that the real market value was far away from the 'hope value' resulting from the portfolio style purchases.

The Project sets out to deal with this in various indirect ways, but also in two particular direct ways which represent 'sticks' and 'carrots'. The 'carrots' are inducing grants to help to bridge the gap between costs and real values including a modest profit for the owner. The costs of residential units in particular, are regulated by the grant sources (EP and SRB) on a 'value for money' basis in order to reduce profiteering and to maintain a modest market growth. The 'stick' is the threat of using Newcastle City Council's statutory powers to compel owners to repair properties or to have them compulsorily purchased at real values, which reflect the unimproved condition of the property and area. The Project and the City Council are currently involved in the complex negotiations that could lead to the compulsory purchase of Grainger Town's long empty cause *célèbre*, 2–12 Grey Street. A careful use of such sticks and carrots can also encourage unsatisfactory owners to sell properties on, which enhances the chances of their proper rescue and considered re-use by a more responsible owner.

Environmental management

Management of transport, parking and urban spaces within the historic environment

As part of the 'Vision for Grainger Town', the area is to become 'a safe and attractive location', and the proper management of city centre transport and the improvement of open spaces and streets have a big part to play in this.

Newcastle's transport policy as described in the 1998 UDP sets out to 'consider radical measures, which can be implemented incrementally to improve the environment and protect the City's long term viability'. This will include removing extraneous traffic from city centre streets, and introducing 'traffic management measures to improve conditions for pedestrians, cyclists and people with disabilities'. The overall intention, particularly in the historic city centre, is to reduce the environmental impact of motorized transport. Finally, measures are to be introduced into the city centre to give 'priority to service vehicles, public transport, cyclists and pedestrians and those private cars essential to its viability'. In short, a civilized transport policy for the twenty-first century.

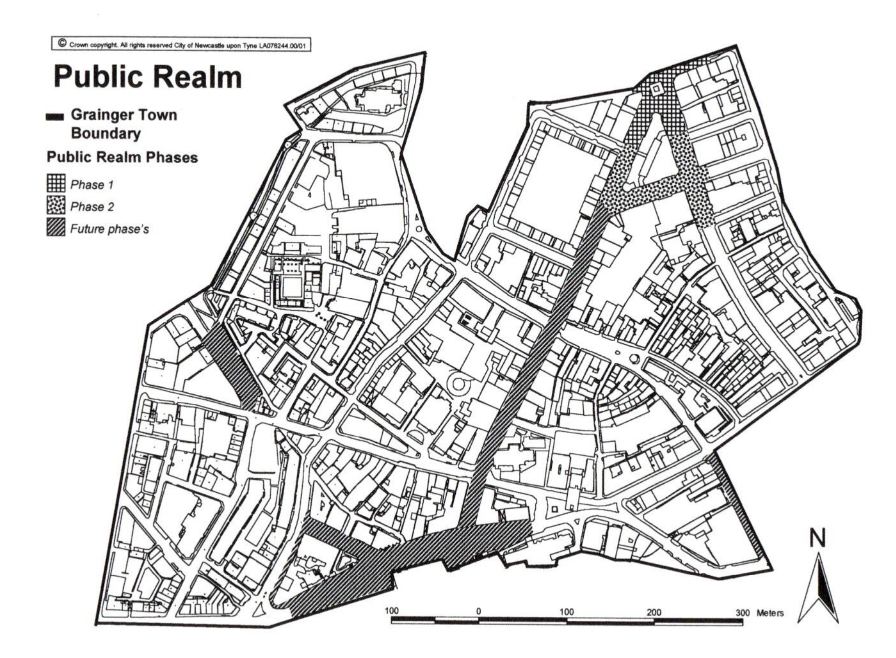

Fig. 13.8 Public Realm Improvement Programme – the Programme's purpose is to improve the main public spaces in Grainger Town using high quality natural stone, and to increase the legibility of the area by upgrading the link between the Central Station and Grey's Monument. (Courtesy of Ordnance Survey.)

There is a strong environmental and quality thread in these policies namely 'improving the environment of . . . conservation areas and other areas of high pedestrian activity', and 'to enhance environmental quality', which makes them particularly applicable in Grainger Town.

The proposed radical measures for the city centre are based on the exclusion of all non-essential traffic from the inner core including historic areas such as Grainger Town. The means of doing this is to designate all roads into the core as 'central roads' on which measures will be developed to discourage and finally exclude unnecessary vehicles. This will result in a single circular route in the core ('the inner distributor road') for permitted traffic, thus enabling the city council and Grainger Town Project to increase the amount of pedestrian dominated space inside the historic core of Newcastle.

As a result of the interactive and broad experience brought to the Project by the Public Realm report (Gillespies, 1998), within the next two years the pedestrian space in the heart of Grainger Town will be extended (Fig. 13.8). Not only will this exclude traffic and extend the pedestrian area, but it will create more high quality urban space for street cafés, public art, creative night lighting schemes and safe perambulations and shopwindow browsing for pedestrians. As in the paving work already completed around Grey's Monument and its surrounding area, the emphasis will be on quality of materials and quality of execution, appropriate to a European Regional Capital of Newcastle's standing. Other areas of street improvement hoped to be completed within the time of the Project include pavement widening and replacing the 1970s landscaping scheme with a more robust and simple scheme.

The prime policy guidance on all parking in the 1998 UDP is that 'provision will be managed to protect environmental quality and the viability of commercial areas, especially the city centre'. The provision of the required 10,000 public car parking spaces in the city centre is to be achieved by allocating land for multi-storey car parks at several locations, three of which will benefit the regeneration of Grainger Town.

The emphasis in the control of development will be upon providing parking only to satisfy operational requirements. Provision proposed in excess of this will be determined by the impact on the environment. Operational provision will be met by implementing parking standards on site, payment of a commuted sum to the city council so that alternative provision can be made elsewhere, or the provision of spaces on a site elsewhere in the locality. Too often in Grainger Town it is impossible to provide the required spaces on site. So the payment of a commuted sum (if not too high) can enable new developments, especially residential, to proceed, thus achieving the re-use of upper floors that is central to the Grainger Town regeneration whilst providing the operational requirements necessary to maintain viability.

Pollution control and policies to limit physical deterioration

The main source of vibration and pollution in Grainger Town, both atmospheric and noise, is traffic. Whilst the project must rely on others to reduce the pollution emissions at source, efforts are being made in Grainger Town to either relocate the traffic sources, or else to screen the more sensitive users from them.

The measures for excluding traffic from the city centre and creating more pedestrian dominant space will help to reduce the air and noise pollution of traffic throughout most of the spaces and streets of Grainger Town. Where higher levels of noise have been recorded and will continue into the future there is a requirement for double or secondary glazing sound insulation in any new noise-sensitive conversions or developments. Grant aid for such insulation is included in any development grant awarded for comprehensive refurbishment under SRB or EP funding.

Tourism and heritage management

The relationship between tourism and heritage management, and the use and conservation of property

A requirement to encourage tourism is implied at the end of the Grainger Town's vision statement in which it is insisted that the area 'will become a distinctive place, a safe and attractive location to work, live and VISIT'. But care must be taken in developing tourism because what appears to be such an attractive objective could turn out to be a 'Pandora's Box' of problems for historic environments.

On the positive side, there is a moral requirement in any civilized society to make its common cultural monuments available to visitors of all kinds, and this can lead to providing a use for old structures as well as generating funds for their conservation too. The wider economy can also benefit considerably from visitor spending. On the other hand, tourism as a land use can be very damaging to the individual historic property or the wider historic environment if it becomes too intensive as happens in the tourist season in other United Kingdom historic centres such as York or Oxford or takes place in an inappropriate location. Popular historic environments may be generally affected but cultural buildings can be particularly at risk from tourism erosion.

The process of developing historic tourist potential must be managed so that the benefits of tourism are maximized while the most damaging effects are avoided. The process is far from perfect as the predication of tourist use and growth is as yet unreliable, but it is much better than accepting unconsidered and unfettered tourist developments. The management process begins with a *Heritage Audit* of the local potential, and proceeds by developing a *Heritage Strategy* by which this potential may be appropriately achieved within the restrictions and timescale of available resources. This strategy would identify city-wide and local themes to be promoted, identify individual buildings or areas to carry this promotion, determine the nature of linking signage, publicity and interpretation and draw up a programme of action and resources capable of bringing this about.

Unfortunately, in the case of Newcastle such a strategy is only in the early stages of preparation at present so there is little guidance, even in principle, available yet to direct heritage and tourist action in Grainger Town. Although certain tourism development decisions that affect Grainger Town have already been made, a comprehensive linking of heritage and tourism will have to await the completion of the strategy.

Even so, the matter of heritage and tourism has not stood still, and the Project and Newcastle City Council have carried out certain relevant activities already. What follows is a list of initiatives taken, some complete, some experimental and some forming a sound basis for future action:

1 Cultural monument improvements
   - developing a scheme to clean and repair the local landmark Grey's Monument;
   - completed feasibility studies for two ancient sites: Blackfriars (a former Dominican friary) and the Castle to secure the fabric and develop their tourism potential;
   - a funding bid under the Europe Union Raphael Programme 1998 to implement the findings of these studies.
2 Tourist environment improvement
   - general upgrading of historic buildings and shopfronts;
   - comprehensive upgrading of space around Grey's Monument as Phase I of a multi-million pound improvement of major Grainger Town spaces (there will be four phases in total);
   - creative lighting scheme for streets and spaces being explored;
   - public art consultants appointed to report on how to develop this aspect of spaces in Grainger Town.

3 Tourist support
- new Tourist Information Centre to open in a shop in Grainger Town in late autumn of 1999;
- put a team of uniformed urban rangers called Newcastle Knights into the city centre and Grainger Town to give help and advice;
- provide Welcome Host training for retail and commercial organizations in Grainger Town;
- provide support for small cultural organizations in Grainger Town;
- developing an economic development strategy to identify and support cultural and heritage industries anywhere in Newcastle.

4 Publications
- two townscape trail books for adults and children developed for Grainger Town by the North East Environmental Education Forum (Lovie, 1997);
- produced postcards of views of Grainger Town;
- prepared an urban heritage handbook (*Investing in Quality*) to raise the investment and tourist profile of Grainger Town's internationally important built heritage;
- publish a newsletter three times a year, aimed at visitors as well as local people.

5 Events
- produced exhibitions, displays, historic plaques, etc., in 1997 to celebrate Richard Grainger's bicentenary (Fig. 13.9) and Year One of the Grainger Town Project;
- produce an annual cultural events programme including street entertainment;
- organized, with the Civic Trust, annual European Heritage Days on which little-known or newly restored heritage structures are open to the public and public tours round Grainger Town area led by Project Officers;
- using the Grainger Town publicity consultants to raise the image of Grainger Town with local and national media.

Fig. 13.9 Richard Grainger plaque – historic plaque commemorating Richard Grainger's residence in Grainger Town. (Photograph: City Repro, Newcastle.)

The cultural role and sense of identity of Grainger Town

The cultural role and a sense of identity of a historic centre are bound up with its distinctiveness. If it has lost its individuality or its uniqueness has become blurred by thoughtless development, then its cultural significance and the identity it gives to the people who originate from the centre, will be diminished or will be replaced by vague feelings of loyalty for what is merely familiar. Having lost its historic distinctiveness, it is still possible for a centre to be given a new unique and memorable image but it will not have the depth and richness of the past that history adds to a place.

Historic centres play various cultural roles: they can provide cultural orientation giving people a sense of belonging somewhere in space and time; the need to 'look back' via historic surroundings gives stability to individuals and identity to local citizens; and the buildings and spaces represent a common 'cultural memory'. Although problems with such common memory may arise as our multicultural society diversifies further, historic centres still represent a significant part of the cultural identity of the United Kingdom at the start of the twenty-first century.

The achievement and maintenance of a unique cultural identity is also built into the end of the Grainger Town vision statement where is boldly states that 'Grainger Town will become a distinctive place'. But, of course, Grainger Town already had a head start in this respect. The fact that it could truthfully be given the name of a single builder, Richard Grainger, and its complex but strong identity can also be given the name *Town*, already indicates that it is a distinctive place, separate from the rest of Newcastle and other parts of Newcastle's centre. It has therefore been one of the objectives of the Project to celebrate, maintain and reinforce this distinctive identity (Fig. 13.10).

Fig. 13.10 Lanterns and music of Richard Grainger's bicentennial children's parade in 1977. (Photograph: City Repro, Newcastle.)

This has been done by the Project in various ways:

1 Celebrate
   - publish reading material describing the common history and qualities of Grainger Town's buildings, layout and principal builder;
   - publish postcard images of its quality details, streets and landmarks;
   - publish a regular newsletter about the aspirations and successes of the many people who have gained from, or been inspired by the area.
2 Maintain
   - repair and bring into use the historic buildings and spaces that characterize the area;
   - specially target the area's landmarks and cultural icons for repair and upgrading.
3 Reinforce
   - improve the public realm to a quality equal to the historic fabric of the area;
   - remove or minimize the eyesores and traffic pressures that diminish or interfere with the historic visual qualities of the area;
   - enhance the historic environment with suitable art adornments and cultural events for all;
   - improve the physical access within the area to avoid excluding members of the community.

Clear confirmation of the cultural success of the project is given by the frequent repetition of the Grainger Town name in newspapers and publications – even children's illustrated books – and its association in these publications with urban quality and distinctiveness.

Sustainability

Historic environment and sustainability

Like all planning and regeneration projects in Newcastle, the Grainger Town Project is currently subject to the requirements of the 1998 Newcastle UDP. This plan includes a section on the desirability of achieving sustainable development. As 'it is clear that the debate on how planning can contribute to sustainability is far from over', the policies in this 1995 plan 'are the city council's first statement on this issue'. These policies for moving towards sustainable development deal with energy and the urban environment, as well as increasing local environmental resources.

Particular sustainable policies actively engaged in by the Grainger Town Project include:

1 Reducing energy consumption and carbon dioxide emissions by:
   - locating the most intensive forms of development at the city centre;
   - retaining areas of mixed use development and activity and encouraging more areas;
   - efficient use of road space;
   - measures to maintain the compact nature of the city.

2 Protecting, managing and promoting the city's built heritage to increase its value and contribution to achieving sustainable development by:
   - reviewing it regularly and extending its protection where appropriate;
   - promoting its proper management in order to conserve and enhance its value;
   - raising public awareness of the value of the city's built heritage and its importance to the concept of sustainable development.

More directly, conservation, as practised throughout the Grainger Town Project, has a special relevance to sustainability since by its very nature it is involved with retaining the embodied energy and resource of existing built assets and seeks their adaptation to new uses. In this respect it will assist in maintaining the vitality of the heritage for future generations, creating a balance between new and old, improving the quality of life by creating a better place to live and to work, empowering community action and responsibility, and, importantly, changing attitudes so that the holistic process of change becomes part of everyone's conscience: property owners, investors, employers and employees, residents, administrators, etc. The present concerns for sustainable development give a new validity to the kind of conservation carried out by such conservation-led regeneration schemes as the Grainger Town Project.

Conclusion

Grainger Town, as a historic urban quarter, has several extraordinary attributes, including:

- its survival intact over 160 years;
- being a remarkable example of nineteenth-century master-planning;
- this plan being overlaid onto a medieval town centre with the fourth most complete defensive walls in the UK;
- comprehensive post-war redevelopment which has had only limited impact on all this quality.

It is the protection and enhancement of these attributes while encouraging total building occupation and environmental improvement that is the fundamental objective of the Project.

Table 13.3 Progress on the number of buildings at risk in 2000 (half-way stage)

	Buildings		*Rescue projects*
1996 (at start of the Project)	Extreme risk	15	–
	Grave risk	14	–
	At risk	36	–
	Total	65	–
2000 (half-way through)	Extreme risk	8	7
	Grave risk	13	1
	At risk	18	18
	Total	39	26

Success in this achievement is measured in objective targets to be attained on a quarterly, annual and Project lifetime basis. These targets cover such things as new jobs secured, areas of floorspace improved, new businesses developed and numbers of new housing units achieved; they also include the number of buildings rescued or improved.

By 2000, approximately half-way through the 6-year programme, all half-way targets have either just been met or exceeded. For the 'Buildings at Risk to be rescued' target, the figures are as shown in Table 13.3.

So at the half-way stage, 40 per cent of the worst buildings have been rescued and re-used as well as 50 per cent of the rest. The balance of 44, i.e. the remaining 39 plus another 5 expected to become at risk within the period of the Project, are programmed for rescue and re-use, using all sources of funds, before the end of the Project in April 2003. This assumes that the current arrangements for spending the large sums available for rescue under SRB and EP will remain unchanged.

In the 'can do' culture of the Board and Project Team, the feeling of achievement is very high. It is reassuring to hear residents and regular visitors alike marvelling at how much has changed in the last few years. The sense of pride in the developing 'café culture' and '24-hour city' image of Newcastle's Grainger Town quarter is becoming more and more evident. Although we still have a distance to go towards our 100 per cent conservation achievement, the Project is well on its way to stablizing the fabric and economy of Grainger Town as well as changing cultural attitudes to the city itself. Both of these successes will have a profound effect for the better in Newcastle well beyond April 2003.

References

City of Newcastle upon Tyne (1987): *Newcastle's Heritage, Planning for Conservation of the City*, City Planning Department.

The Conservation Practice and others (1992): *The Grainger Town Study*, Commissioned by Newcastle City Council, English Heritage and Department of the Environment.

EDAW (Planning and Economic Development Consultants) (1996): *Grainger Town Regeneration Strategy,* Newcastle City Council and English Partnerships.

English Heritage with Town Centres Ltd and the London School of Economics (1999): *The Heritage Dividend.*

Gillespies (Planning and Design Consultants) (1998): *Grainger Town – Vol. III: Public Realm Strategy*.

Lovie, D. (1995): 'Newcastle upon Tyne: The Grainger Town Project', in Burman, P., Pickard, R. and Taylor, S. (eds) *The Economics of Architectural Conservation*, Proceedings of a Consultation, 13–14 February 1995, Institute of Advanced Architectural Studies, University of York.

Lovie, D. (1997): *The Buildings of Grainger Town: Four Townscape Walks Around Newcastle,* NEEEF, Newcastle upon Tyne.

Newcastle City Council and English Partnerships (1997): *Grainger Town – Working for a New Living*.

Urban and Economic Development Group (URBED) (2000): *Investing in Quality*.

Robert Pickard and Mikhäel de Thyse

The management of historic centres: *towards a common goal*

Introduction

Since November 1998 'rehabilitation' and the role of management strategies in historic centres has become the subject of debate and analysis by the Council of Europe's technical co-operation and consultancy programme in the field of cultural heritage. This examination of different historic centres is in parallel to the programme's aims and provides evidence in the same context.

The current debate has centred not just on heritage protection but also on the wider issues of management involving both the community and policy-makers and encompassing the questions of social cohesion and economic development, on the need for functional diversity and the importance of identity. From this three key themes have emerged: Urban Values, Political and Institutional Frameworks, and Intervention Methods are all regarded as being important in the management process (and will be considered in relation to this sample of centres investigated).

The debate has brought out and confirmed a number of situations and issues, often similar but sometimes peculiar and distinct, which prevail in several European cities with regard to the development of their historic centres:

1 An increasing trend towards depopulation (and sometimes the loss of the residential function).
   - The historic centres in Malta have suffered dramatic population abandonment – partly encouraged by relocation into social housing schemes on the urban periphery rather than rehabilitating existing older houses. The population of the historic centre of Venice has reduced by over 100,000 in the last 50 years and the trend is continuing.

2 General deterioration of the older areas, leading to undesirable changes that deprive centres of their character and destroy the social fabric that makes up a city's identity and fosters such values as solidarity, belonging and social control.
   - In Zemo Kala (Tbilisi) the groundwater problems, lack of maintenance and scarce resources of the residents is a real threat to this strongly bound community – made worse by development pressures that are not subject to effective control.

3 A recent urban history marked by the development of initiatives on the part of the community.
   - For example, the Dublin Crisis Conference in 1986 brought together a wide range of conservation, community and environmental groups that called for official action for the sensitive renewal of the inner city. This was the starting point for concentrated action.

4 An increasingly uneven development of real estate markets as property prices rise with the development of service activities and economic activities in general.
   - In the boom years of the 1960s and 1970s speculative property investment in the United Kingdom had a detrimental impact on centres such as the historic core of Newcastle (Grainger Town) as 'hope value' was not realized. Similarly the historic centre of Santiago de Compostela suffered from a tertiary boom that pushed the resident population out. In developing economies such as the Czech Republic and Latvia, market forces are now causing some concerns in this respect – emphasizing the need for proper management tools for historic centres.

5 Large-scale deterioration in the condition of historic buildings and increasing numbers of empty properties or spaces within them.
   - In Grainger Town, where there is one of the highest densities of listed buildings in the United Kingdom, the survey of buildings at risk due to disrepair or vacancy carried out in 1991 identified that 47 per cent were at risk (compared to the national average of 7 per cent). Moreover, in Erfurt over 300 endangered buildings were examined at the commencement of action (also in 1991) to halt decay and including clarification of ownership – a particular problem for historic built environments of the former totalitarian regimes of central and eastern Europe.

6 Demolition which could be avoided but, instead, breaks up the structure of the city and signifies a major waste of materials and energy, all in the name of modernity and progress.
   - The Grand Harbour area in Malta has seen repeated calls to demolish seventeenth- and eighteenth-century housing due to their poor condition. Also in the Temple Bar area of Dublin terraces of eighteenth-century buildings were demolished in the early 1990s to make way for new architecture.

7 Incompatibility between the classic urban reinvestment patterns (growth of public or private service activities) and the desire to leave residents in place (represented in some cases by the loss of the residential function).

- The booming economy of Ireland, and in particular Dublin, has focused the debate on the historic built environment and on the community – as to whether they will be sufficiently safeguarded in the pace for change.

8 Functional and social segregation, destroying the diversity that constitutes the very essence of a city.

- The centre of Valetta has experienced a steady socio-cultural decline due to a domination of business and commerce and the secondary consideration of community activities including the relocation of residents to outer areas.

Moreover, in some centres the encouragement of tourism without proper management has had a significant and detrimental impact upon the physical and social fabric – as is evidenced in Venice.

The debate has helped to pinpoint certain difficulties that many centres have in common when it comes to adapting urban strategies to contend with new situations:

- the relative instability of urban management structures in historic centres;
- the increasing demand for improved heritage protection instruments (plans, financing);
- decreasing or inadequate long-term planning tools;
- failure of public means of action to keep pace with growing and more varied needs;
- a strong tendency to combine public and private financing as a matter of course to fund programmes which are too selective (office developments, shops, up-market housing) and which yield rapid returns – rather than taking an holistic approach;
- difficulty in bringing together the different levels of public solidarity at a given time (a lack of inter-municipal structures and limited financial resources at municipal level, and a lack of state funding);
- increasing ineffectiveness of traditional consultative machinery and difficulty in organizing any form of permanent dialogue with local residents' and business representatives.

This has led to the need for new active methods to enable a new role for historic centres, through an integrated management process, in line with the social, heritage-related and urban issues at stake. Indeed, focus is now being directed towards:

- maintaining the social links that bind the community together;
- preserving the residential function of old city centres and to revive it where it has been lost;
- preserving the economic, social and cultural functions of old towns and cities and ensuring there is a balance of functions;
- helping to improve environmental conditions in old centres;
- conserving as much of the built heritage as possible in a sustainable way by encouraging rehabilitation;

- Defending the identifiable, human scale in a world increasingly dominated by networks and constructions that destroy and deny human contact and generally breed alienation.

Voluntary and long-term integrated activities, involving local authorities, other public authorities, the economic players and the population, have to be carried out by mobilizing means commensurate with the needs of each centre.

Action

The different types of action call for a combination of rules and conditions to serve as basic principles and factors for success. These must be based on safeguarding recognizable urban values, on ensuring the political commitment and the development of an appropriate management framework, and on formulating suitable intervention methods and implementing them through the management process.

Urban values

Identity and *diversity* are perhaps the most important aspects of urban value as reflected in the following issues.

The recording of the physical characteristics and condition of the historic environment (the architecture and townscape features: street layouts and finishes, parks, 'green' areas, landscaping and other monuments and features) is an important part in this process of recognizing and safeguarding the sense of belonging and *identity*. This assessment should form the basis for developing physical preservation and enhancement plans and historic building revitalization and rehabilitation. Moreover, the essential urban design characteristics with which people identify can be strengthened through the formulation of advice and guidance and through development briefs for sites that have a negative or neutral impact within a centre. The question of *identity* can be looked at in relation to the following three issues: (i) respect for the morphology and typology; (ii) the importance of public areas; and (iii) the perceeption of architecture.

Respect for the morphology and typology

The morphology and typology of historic centres must be respected and rehabilitated in accordance with certain architectural and technical stipulations (for example, this was the starting point for the Rochefort architectural charter – the basis for land-use regulation). Respect does not necessarily mean conserving a building in its entirety but reviving its spirit and identity. It means being flexible enough to adapt the objectives of rehabilitation to the needs of modern living while respecting the values generally upheld by the local community. Once rehabilitated along these lines, the area should 'recognize itself'.

The importance of public areas

Rehabilitation should address habitat as well as buildings. Public areas are of the utmost importance, essential as they are to the quality of life in a neighbourhood and to the way in which people perceive and identify with their habitat. It is therefore paramount that projects include public areas selected for their power to strengthen people's sense of belonging. The areas concerned thus become bonding factors between the citizens and their habitat.

The perception of architecture

Whilst respecting the morphology and typology of the historic centre, just as they identify with the public areas, residents must identify with the architecture in their neighbourhoods. In so doing, they become its principal defenders. But they will identify with it only if it remains authentic, only if it is not artificial and out of touch with reality, if it is consistent with their own values and experience. Examples of how these issues have been dealt with can be identified as follows:

- In Bruges, 'identity' has been strengthened by the renovation of squares that have become new meeting places for inhabitants and have been used for city functions, as well as providing the incentive for the restoration of surrounding buildings. There has been a significant improvement in streets and modernization of sewers and canals. Furthermore, an aesthetics committee, design-guidance commission and city regulations have all had an important influence on design matters.
- The aim of the architectural charter developed for Rochefort was to express the identity of the town. Apart from its use as a planning and rehabilitation guide, it was directed at the public to explain concepts of urban form based on an analysis of the morphology and typology of the urban heritage. Moreover, at the centre of the project to revive the town of Rochefort, the notion of public space has become a fundamental consideration.
- In Grainger Town, cultural 'identity ' is central to the regeneration philosophy. This can be seen through the encouragement of good design in rehabilitation works (such as the assistance given to renewing shopfronts in harmony with the buildings that they front and in the public realm initiative for the townscape environmental improvement). But more particularly by fostering the sense of belonging – identity to local citizens is emphasized through the connection with the past (by the name of Grainger) and by a process of community celebration, cultural events and awareness raising.
- In Erfurt, the traditional courtyards form part of the integrated policy of preserving historic spaces, thereby aimed at improving the quality of urban life – moreover, public spaces are linked to parks and to mixed use accommodation for residential and business purposes.
- In Tbilisi, the multicultural and multi-ethnic identity of the Zemo Kala pilot site remains today. The importance of safeguarding this mixed community is an essential part of the process of rehabilitation. This is why *all* the buildings in the area have been recorded, as well as the urban fabric, townscape and street network – as a preliminary move to developing integrated management tools for the rehabilitation project.

- In Ribe, particular attention has been paid to historical and topographical conditions including distinctive building patterns and characteristics of individual periods through the Survey of Architectural Values in the Environment (SAVE – Preservation Atlas) and through guidance to ensure historic design traditions are safeguarded.
- In Telč and Riga, action has commenced to survey, record and then develop databases (and to provide instructions for works on existing buildings in Riga) as a basic structure for analysing the identifiable important elements and character of the built heritage that should be safeguarded through planning tools.

Diversity in both a social and functional sense (the two being inter-linked) helps to breed harmony and vitality. The relationship between commercial and social needs must be carefully balanced.

Social diversity
The social diversity that is generally still found in old city centres must be maintained and even encouraged. 'Ghetto' formation and gentrification are two phenomena with similar effects (as noted in Malta), powered by the same forces, which must be fought with vigilance and determination. Social diversity is a sign of stability, of a healthy, thriving urban environment.

Functional diversity
Like social diversity, functional diversity has always been a feature of old centres. It too must be preserved – moreover the ability to allow change should be preserved. The constraints and pollution such diversity may have caused over the years must be corrected, of course, but functional variety should be maintained, the different functions co-existing in a delicate but necessary equilibrium based on new criteria of compatibility. Where appropriate (i.e. where there is flexibility to allow change in a physical sense of change) this co-existence can be encouraged through the permitting of mixed uses within individual buildings.

In some countries social and functional diversity has been safeguarded. For example:

- The old quarter of Erfurt professes a strong mixture of business and residential accommodation. This has been ensured by political commitment – official policy has stated that modern forms of shopping, service and recreation are not to disturb the existing balance.
- The Grainger Town Project has placed a strong emphasis on social and functional diversity through the concept of 'Working for a New Living': to increase the residential population (both for affordable housing and for sale); to improve training and employment opportunities; to safeguard and promote business interests; to promote arts, culture and tourism; and to rehabilitate historic buildings for housing, offices, shops, arts and leisure uses.
- The systematic rehabilitation and re-use of buildings in Rochefort has shown this effect – by attracting large private enterprises and the conse-

quent benefits for employment in the town and improving housing at the same time.

On the other hand:

- There is still concern in Dublin, due to the strength of the economy and escalating property prices, that current rehabilitation action could threaten the cultural continuity of neighbourhoods. In the Smithfield Village for instance, traditional markets may be lost at the expense of innovation. At the same time, the action in support of the Temple Bar Project arose and built upon the unique mix of cultural activities and small business that developed there once the threat of demolition had gone.

Political and institutional framework

The management of historic centres requires a political commitment both at national level, through financial support mechanisms and regulatory and legislative provisions, and at the local level, where municipal action can be directed in association or in partnership with other agencies and with the private sector through an integrated process.

Political commitment

Firm, unfaltering political commitment is an absolute necessity in any urban rehabilitation project. It has a direct effect on the acceptability of the project to the local population and on the motivation of the teams concerned. If the operation in terms of its overall objectives ties in with broader schemes of planning, its chances of success are all the greater. The commitment must be constantly renewed, for these are long-term operations, even if politicians are usually more at home with short-term objectives. The commitment must also be reflected in a suitably flexible framework for public authority action to be updated as the work progresses. The resulting system of rules and rights is in the realm of regulatory or legislative codification governing housing, construction, urban planning, heritage and finance. However, the importance of developing practical means of partnership with the inhabitants and business interests is also important in this context.

- The evidence of Santiago de Compostela highlights the fact that the process of securing a political agreement can be an arduous process, fraught with many difficulties, but nevertheless worthwhile. Citizens have been introduced to the advantages of rehabilitation, and the damage to the built heritage, incurred by normal patterns of property investment, has been curtailed.
- The Grainger Town Project further shows the relevance of partnership and commitment by the local authority with other national and local agencies and partners: funding, conservation, cultural, training, business and regeneration agencies, as well as other representatives of local interests.

For developing countries, in a process of economic transition, this process may be more difficult. National and local political structures need to be adaptable to the needs of the community and the demands of democratic involvement.

- For the historic centres of Riga and Telč the process is just beginning with international assistance from the Council of Europe and other partners. In the former a co-ordination board has been established – signalling a commitment to action.

There is therefore an important role that management agencies (public, and in joint venture with the private sector) can play in co-ordinating the rehabilitation and revitalization process.

Political commitment can be also in the form of 'sticks' (such as the taxation of long-standing vacant buildings in Bruges and the serving of statutory repairs notices in Grainger Town) and 'carrots' (incentives). The possibilities for financial incentives vary according to the resources available in each country and each locality. For example:

- In France the national agency, ANAH, plays an important role in association with municipal authorities – this combination with the OPAH mechanism has had a dramatic effect of Rochefort.
- Before the Grainger Town Project was established there was a distinct shortage of funding opportunities. But action is now supported through partnership with regeneration and training agencies as well as specific conservation funding directly from the municipality and with the national agency for the historic environment though a specific Heritage Economic Regeneration funding scheme.
- For countries like Georgia these schemes are just not possible, but international assistance is available if the authorities wish to make the commitment to action – as evidenced by the support of the World Bank in Tbilisi. Moreover, the support has been directed to the vulnerable sectors of society with a view to safeguarding the community of the area.
- However, as can be noted from the example of Dublin, the setting up of an urban renewal agency and the provision of financial incentives may not have the desired effect – unless the aims and regulatory procedures are sufficiently directed to the goal of conservation and the safeguarding of cultural identity.

The problems that now exist in Venice reveal the dangers of not having a firm political commitment to the importance of managing historic centres:

- Depopulation has occurred, social and functional diversity has been lost, severe environmental problems exist and excessive tourism has put the survival of the city at risk. The planning process, spread over a 30-year period, did not result in effective planning tools. It is only recently (since 1992) that more effective regulations have been being introduced and even these have required several amendments. However, there has been considerable expenditure in an attempt to arrest the problems, if not improve the situation.

Rehabilitation as a component of official housing policy

The status of old housing stock on the local housing market is well known. In order to aid and support the rehabilitation of dwellings (and other buildings for residential use), housing policy instruments must be adapted to household income and the specific constraints of the built environment. They should target a social housing function and the maintenance *in situ* or re-housing of residents in lower income groups. In the context where the residential community has been depopulated it will be important to ensure that that this process is encouraged with the aim of creating a balanced community for the future, rather than rehabilitating by a process of gentrification that serves an exclusive few:

- While poor social conditions in Malta resulted in social housing developments, the community has been displaced and an opportunity was lost in terms of existing dwellings, although this may now be corrected due to the greater attention being paid to rehabilitation.
- In Dublin, although there has been substantial gentrification within the historic core, recent action by the corporation seems to have turned the tide – through direct action to purchase flats to protect a socially mixed neighbourhood.
- In Bruges, there has been a long-standing commitment to maintain the number of council dwellings and to rent some houses at a social rate. Moreover, a public commission for social welfare has ensured cheap accommodation for pensioners in historical complexes that have been systematically restored and renovated in recent years.
- The rehabilitation of empty spaces (for example, in the upper floors of commercial premises) has helped to ensure a mix of uses – an example also utilized in Grainger Town, which has been encouraged by the promotion (financially and otherwise) of the concept of 'living over the shop', and also in Rochefort, through a recent OPAH programme.
- In Rochefort, the combined effects of actions by the national housing improvement agency (ANAH), the municipality and the use of the funding programme (OPAH) has played a significant role in rehabilitating older housing.
- In Santiago de Compostela the action on housing has ensured that over 300 dwellings have been rehabilitated, with and without subsidies, for social housing and otherwise, to help ensure a proper social balance.

An integrated approach to planning

Nowadays cities are faced with the phenomenon of world-wide networks and the resulting dilemma between the gigantic scale of such networks and the need to revitalize communities on a local scale. The notion of neighbourhood identity, one of the aims of urban rehabilitation policy, is not limited to city centres. When rehabilitation work is being planned, attention must be paid to the situation of the built environment, the community and their needs and the impact on the immediate neighbourhood, the district and the city as a whole (re-emphasizing the need to survey the state and condition of the historic environment). The work should be part of an urban process designed to improve physical conditions and, where appropriate, revitalize the local economy, for example by maintaining local services and reorganizing traffic and transport to alleviate car-related nuisance (see the section below, 'environmental management').

In some countries the integrated process of planning is in its infancy:

- The proposed rehabilitation/preservation plan for the historic centre of Riga is expected to play an important role for this centre in the future by bringing together the different interests and bureaucracies towards a common goal. Its model may be used elsewhere in Latvia for other historic centres in the future.
- In Telč the process of developing integrated methods has started – by drawing on good practice examples from elsewhere.
- New planning legislation is currently being drafted in Georgia and reflects the importance of integrated conservation methods, but it will be some time before practical policies can be fully implemented in Old Tbilisi.

Elsewhere, specific planning tools have been adopted to ensure an integrated approach to historic centre management. For example:

- In Bruges, the innovative system of Master Plans has been devoted to comprehensive examination of the interaction between the historic city and the surrounding region in which town planning must contribute to the socio-economical and cultural development – and the model has been adopted in other historic towns in Flanders.
- In Rochefort, the architectural charter has served as a basis for land-use planning, following very closely the principles of the 1987 ICOMOS Washington Charter.
- In Santiago de Compostela, the guidelines of the Special Protection and Rehabilitation Plan have been regarded as an optimum instrument for organizing rehabilitation and ensuring that a proper balance of social, commercial and other functions have been integrated in this historic centre.
- The review of the city's heritage served as a basis for integrating conservation policy into the unitary (strategic and detailed) development plan for Newcastle – a policy framework (reflecting all land-use issues) for the Grainger Town central area.

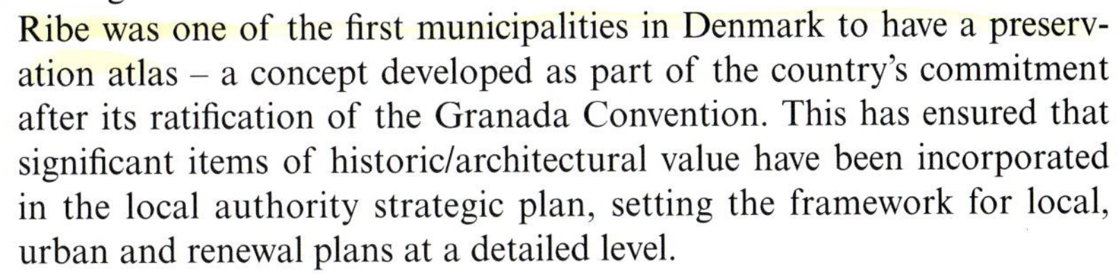

- Ribe was one of the first municipalities in Denmark to have a preservation atlas – a concept developed as part of the country's commitment after its ratification of the Granada Convention. This has ensured that significant items of historic/architectural value have been incorporated in the local authority strategic plan, setting the framework for local, urban and renewal plans at a detailed level.

Striking a balance between public and private sectors
The correct balance must be struck between public and private action. Rehabilitation and regeneration should not be exclusively public or private, especially when it comes to preserving social and functional diversity. In the initial stages, of course, many projects are mainly public, the private sector becoming involved only when the prospect of making a profit begins to materialize. In the later stages, private participation should be predominant, but with some participation by the public sector to preserve the initial social goals. In an old urban environment, subsidized housing will always be necessary, as will public contributions to the maintenance of social facilities, as well as the need to safeguard commercial interests.

The partnership approach ensures that community needs can be safeguarded and with the assistance of the private sector, although the private sector will have other aims (but these may also be of public benefit).

- The Grainger Town Project was commenced on the basis of a £120 million scheme – the proposal was for £40 million to come from the public sector and the remainder to be attracted from the private sector. So far the evidence suggests that this type of partnership approach is working, with rehabilitation action attracting more private finance – four or five times the initial public investment.

Management tools and intervention methods
The implementation of strategies for the management of historic centres requires appropriate management organization and intervention methods and participation by the community itself.

Keep the project on a manageable scale
Generally speaking, the results to be achieved are quite substantial, in terms of both quantity and quality. Historic centres can cover large areas and tend to accumulate several urban problems. Planners must take care not to 'bite off more than they can chew'. Projects should be kept to a 'manageable' level; they should be high-profile and feasible within a reasonable time and at a reasonable cost if they are to be credible in the eyes of decision-makers and those responsible for their implementation. They should rapidly produce a measurable knock-on effect and help to build up momentum.

Careful planning and organization of strategies is important in this context. The setting of specific management boards or quasi-public agencies in association with the municipal authorities may be important in order to implement a strategy and co-ordinate actions on the ground, and to provide links to the other authorities and agencies, business groups and community associations.

- The setting up of the Grainger Town Partnership Company and Board with direct links to the local authority and consultation and advisory groups has been a very effective vehicle for the rehabilitation and regeneration action in this centre. But this is a large city centre – operations on this scale cannot be utilized in every historic centre. Nevertheless, other examples follow a similar pattern (for instance, in relation to Santiago de Compostela where the City Consortium established a decentralized rehabilitation team; see below)

Interpretation as a factor of appropriation

Familiarity and information play a major role in the mechanism of appropriation, which is itself an essential part of the process. The more familiar people are with their historic environment and the more aware of its value, the more they identify with it and appropriate it. Familiarizing them with their surroundings by various means must therefore be a part of any management strategy: leaflets, lectures, radio and television programmes, posters, discussion groups, and within the education process (schools, colleges, etc.). A rehabilitated building with which the community identifies strongly becomes instrumental in its own rehabilitation. In other words, the implementation strategy must involve a process of 'awareness-raising' and encourage interpretation and active involvement by the community: it is important for people to feel that they are gaining an advantage through the enhancement of the heritage.

- In Grainger Town, awareness-raising has been encouraged through a process of celebratory involvement and backed up by regular publications to keep the community informed of progress in the regeneration project and other material to describe and illustrate the area.

Democratic management

Considering the varied nature of the problems posed by urban rehabilitation, particularly its effect on the functions of a district and the social and property-related changes which it may cause, it is altogether legitimate for the public to have a hand in determining policy choices. Moreover, the 'active' democratic participation by the wider community (representing inhabitants and commercial interests), including involvement in solving day-to-day problems in the life of the historic centre should be encouraged.

Neighbourhood, voluntary and amenity associations are protagonists more than relays in the process, indeed its mainsprings in most cases. The machinery of participation by both the resident and the business population must be organic in order for the project to become a collective one. This democratic imperative requires a personalized style of management involving multi-disciplinary local teams working together in a co-ordinated manner.

- Significantly the commercial regeneration activity within Victoria (Goza – one of the Maltese islands) has been encouraged through the combined actions of the local council and local business interests – multiple small-scale private investors.
- In Dublin, after considerable review of the inadequate protection measures for historic buildings, there is now a greater acknowledgement of the

importance of co-operation with voluntary conservation bodies in reviewing buildings for listing through development plan objectives.

Decentralizing intervention teams
In the interests of democracy and implementing projects on a manageable scale, it is important to decentralize work and have it carried out by local agents in direct contact with the community. The local structure thus formed should be given maximum responsibility to permit an integrated rather than a sectoral approach. This requires an interdisciplinary team, whose varied skills are a source of mutual enrichment. Moreover, as built-up areas are complex environments, effective action requires different interacting skills, none of which, singly, holds the key to success. This is why multi-disciplinary teams have become essential.

- On a small scale this can be seen in the example of Rochefort where a team comprising officials from different disciplines in municipal services, with the assistance of outside specialists, has been closely involved in the renaissance of the town.
- For the medium sized city of Santiago de Compostela, by political consensus, the need for a management structure (the City Consortium) was defined, through which a specific subordinate multi-disciplinary 'rehabilitation office' was set up.
- In Erfurt the approach has been to establish a 'redevelopment agency', decentralized from the local authority, which acts as a trustee for the administration of funds (as is the case for the *FUND* in Georgia).
- By contrast, for the larger city of Newcastle a specific company with its own director and board was established for the Grainger Town Project.

In developing countries it may be necessary for action to be taken at the national level:

- The Fund for the Preservation of Cultural Heritage of Georgia is an independent entity but works with local and national authorities and international bodies. It has concentrated action on a number of pilot projects including the historic centre of Tbilisi and the historic town of Signagi.

A two-way strategy
In this type of operation, a two-way strategy – 'top-down and bottom-up' – must be adopted for optimum efficacy. Since planning is essentially a top-down process, if the need for decentralization is to be met and the people most directly affected – namely local residents and shopkeepers and other local businesses – are to have any say in the matter and influence the course of events, a bottom-up approach is also required. The planning process must be flexible enough to incorporate the demands voiced at grass-roots level.

In this respect action by community interests change the opinions of the policy-makers and implementers as evidenced in many different causes – for example in Dublin:

- The Dublin Crisis Conference was created by voluntary interest groups of students and conservationists, who argued for saving the Temple Bar area from destruction, and other groups that have been critical of policy-makers. But community groups in Dublin are now being given a greater say at the earliest stages of proposals, reducing the need for vociferous action due to a new a spirit of co-operation and involvement.

Environmental management

To reduce costs and satisfy needs to the greatest possible extent, to avoid wasting materials and energy, to limit the amount of waste produced and the inconvenience of transporting heavy materials, and for the sake of authenticity, as much as possible of the existing structures should be conserved. This means carrying out maintenance work and repairs, and restoring using traditional techniques and materials, where possible produced locally by methods compatible with sustainable resource management and renewal. Skilled workmen and technicians who are familiar with these techniques must therefore be utilized, which in turn can help to provide employment in the centre.

- Specific attention has been made to design and guidance on renovation, maintenance, repair and restoration action in Ribe to ensure that work on historic buildings respects authenticity.
- In Bruges and Grainger Town, both work on existing buildings and new developments in the historic environment are subject to independent analysis.
- In Malta's historic centres there has been a problem of the gentrification of older houses (and social exclusion), with alterations being allowed without any notion of conservation standards and authenticity, although the strengthening of conservation regulations and the establishment of rehabilitation committees in recent years should assist in reversing this situation.
- In Dublin, the Temple Bar quarter has also been gentrified. The area has allowed change, as evidenced by new and exciting contemporary architecture, but this has been at the expense of existing buildings – many of which have lost their essential characteristics – a situation that should now change due to the strengthening of protection measures in 2000.

Apart from the environmental context of buildings, the authenticity of the townscape should be safeguarded and enhanced. Environmental erosion caused by heavy traffic, inappropriate transport management measures and consequent pollution can have a significant impact on the historic centre. Opportunities to reduce these and to reclaim the environment for the public by improving the physical fabric of streets and street surfaces, the use of pedestrian priority schemes and other features such as landscaping, the creation of public spaces and the planning of parking areas, will need to be considered. Moreover, this requires co-ordination with other responsible agencies and authorities.

Local councils have been the driving-force in introducing better traffic control and parking regulation – often resisted at first – but now the improvement of the environment is well accepted by the community and indeed expected:

- In Bruges, traffic-reduced pedestrian areas have been pursued and other measures such as a loop traffic circulation system, underground car parking, reduced speed areas and special buses have encouraged a friendlier environment and removed nuisance factors.
- In Grainger Town, the on going 'Public Realm' project has already created a pedestrian friendly environment and traffic management measures are gradually moving towards the exclusion of all non-essential traffic in the inner core.
- Similar action has taken place in Santiago de Compostela, Ribe and Erfurt – in the latter, private traffic within the historic core has been at reduced levels for decades, at the same time safeguarding access by car parking provision on the edge of the centre.

Tourism management

Tourism can be economically beneficial to historic centres, creating financial benefits and supporting local businesses. But it must also be balanced against the needs and desires of the existing community as an over-emphasis on the tourist function creates pressures for new services and associated development, sometimes to the detriment of the local population, and can to lead to damage to cultural assets. It is therefore important for the tourism capacity to be carefully managed and controlled in a sustainable manner:

- In Bruges, which attracts some 3 million tourists per year, the damage caused by tourism has been well recognized as evidenced by the community taking action itself to ensure a bearable living style for residents. The protest movement led to specific action – concentrating tourist activity in the 'golden triangle' to limit the impact elsewhere and through building licence control to restrict the number of lace and chocolate shops, and hotel accommodation, etc., in order to safeguard essential services for the community.
- By contrast, in Temple Bar (Dublin), the marketing of the area as a tourist attraction, while being important to the revival of this historic quarter, has had a negative impact on important heritage buildings and has been viewed as a 'shopping and dining Mecca for the affluent few'.
- There is already a warning that increasing tourism could have a detrimental impact in Santiago de Compostela, but for the moment matters are cautiously being observed and kept in check, and a specific tourist plan has recently been developed to assist this purpose. Similarly, for the city of Telč, the need to manage increasing tourist activity is recognized.
- The problems of mass tourism in Venice are well known and now difficult to manage. Tourism has had a severe impact on the whole life of the city – reducing the service provision for local residents, causing

congestion and many other problems. This, at least, provides a lesson for other historic centres – that tourism should be a sustainable activity.

Towards a common policy: the need for sustainable methods of management

In the light of the experience of the European cities, further reflection must take – as its starting point – the idea that effective action to promote the management and rehabilitation of historic centres necessitates the adoption of a common and holistic policy based on the following assumptions:

- Historic centres are meaningful only if inhabited; they are the expression of the lives of the people who live and work there.
- Free-market forces must be kept in check to prevent cities and neighbourhoods from losing their identities – a tendency that is exacerbated by the processes of pauperization and gentrification.
- Old city centres must function in such a way as to improve the conditions for inhabitants and reduce the effects of poverty.

The common policy to which these assumptions give rise must take into account four principles that determine the approaches to be adopted:

1. Conservation of the built heritage, regardless of its monumental value, as a framework for urban projects and urban management and as an expression of sustainable development.
2. Improvement of neighbourhood amenities: action on housing, public areas and environmental amenities (façades, streets, squares, markets and the wider townscape) which contribute to the feeling of belonging.
3. Strengthening of the notion of centrality: upgrading the various central functions (shops, housing, neighbourhood and district amenities) – a balanced mix of functions helps to maintain vitality and harmony.
4. Action to promote economic activity: employment, tourism, involvement by the local population.

Three main lines of action are identified as vectors and support for activities:

1. Rehabilitation and construction in city centres to help develop social diversity.
2. Enhancement and promotion of public areas as part of the local habitat, areas with which the local community can identify.
3. Democratic debate as a way of legitimizing the action taken and giving it a meaning.

The aim should be to rationalize the past with the present and future needs of community. Historic centres provide an inheritance of the past and the product of change that society has required for its needs. This evolutionary process should continue so that society today leaves its own mark on the environment for future generations. The most important issue is to look to the long term and plan the balance of permissible change and ensure the rehabilitation and revitalization of the historic centre in a sustainable manner.

However, the concept of sustainability is still under scrutiny. Further analysis is required for the formulation of heritage management methodology within a framework of sustainable development. For some countries, and individual centres, it is clear that there is sustainable practice but little evidence of *specific* policies in this context. At the same time, evidence from this study, in particular in relation to Dublin, Newcastle and Santiago de Compostela, has shown that 'sustainable development' is becoming a key factor in strategies for the management of historic centres.

Agenda 21 – as defined at the United Nations Conference on Environment and Development (the Earth Summit) held in Rio de Janeiro in 1992 – calls for 'indicators of sustainable development' as a tool to assist the process of moving towards a more sustainable society. Such indicators are not meant to be precise but may be used by decision-makers as a means to assess developing trends, pressure points or the impact of policies that have been formulated and put into action. They are meant to highlight policy options and other methods of providing responses to change. No precise indicators have been developed in relation to the built heritage. Moreover, there is no right solution and this process of developing indicators within a framework of sustainability will inevitably involve trial and error.

However, some general principles may be identified that are relevant to the sustainable management of historic centres:

- respect community life;
- improve the quality of life;
- maintain identity, diversity and vitality;
- minimize the depletion of non-renewable heritage assets;
- change attitudes and perceptions – the process of managing change involves wider interests and should involve different actors from the public and private sectors; property owners, investors, residents, and other community and voluntary interests. In other words, the process should become part of everyone's conscience;
- empower community action and responsibility through involvement;
- provide a suitable policy framework for integrating conservation objectives with the aims of sustainable development;
- define the capacity by which the historic centre can permit change.

In order for the built heritage to be included in the aim of building a sustainable society, the 'static' goal to protect must be married to the managed 'process' of change within a community framework of planning and negotiation. The capacity of the historic centre to accept change will depend on relative values that are placed on heritage assets and the priorities of society.

Index